INCOMPATIBILITY IN FUNGI

A Symposium held at the 10th International Congress of Botany
at Edinburgh, August 1964

Edited by

KARL ESSER
Ruhr-Universität Bochum (Germany)

JOHN R. RAPER
Harvard University, Cambridge, Mass. (USA)

SPRINGER-VERLAG · NEW YORK INC.
1965

ISBN 978-3-540-03334-9   ISBN 978-3-642-87052-1 (eBook)
DOI 10.1007/978-3-642-87052-1

Contents

Contents

Contributors

AHMAD, MAJEED, Ph. D., Professor of Botany, University of Dacca, Department of Botany, Ramna, Dacca 2 (East-Pakistan)

BISTIS, GEORGE N., Ph. D., Assistant Professor of Biology, City College of New York, Department of Biology, New York, N. Y. (USA)

BURNETT, JOHN H., Ph. D., Professor of Botany, University of Newcastle-upon-Tyne, Department of Botany, Newcastle upon Tyne 1 (England)

DAY, PETER R., Ph. D., Chief of Genetics Department, Connecticut Agricultural Experiment Station, New Haven, Conn. (USA)

DICK, STANLEY, Ph. D., Assistant Professor of Botany, Indiana University, Department of Botany, Bloomington, Ind. (USA)

ELLINGBOE, ALBERT H., Ph. D., Assistant Professor of Botany and Plant Pathology, Michigan State University, Department of Botany and Plant Pathology, East Lansing, Michigan (USA)

ESSER, KARL, Dr. phil., o. Professor, Institut für Allgemeine Botanik der Ruhr-Universität Bochum, Bochum-Querenburg, Im Lottental (Germany)

MATHER, KENNETH, Ph. D., Professor of Genetics, University of Birmingham, Department of Genetics, Birmingham 15 (England)

PARAG, YAIR, Ph. D., Lecturer in Botany, Hebrew University, Department of Botany, Jerusalem (Israel)

PRUD'HOMME, NICOLE, Maître-Assistante Laboratoire de Génétique Physiologique du Centre National de la Recherche Scientifique, Gif/Yvette (S.-et-O.), (France)

RAPER, JOHN R., Ph. D., Professor of Botany, Harvard University, Biological Laboratories, 16 Divinity Avenue, Cambridge, Mass. (USA)

SNIDER, PHILIP J., Ph. D., Associate Professor of Biology, University of Houston, Department of Biology, Houston, Texas (USA)

Contributors

Foreword

Sexual reproduction in the fungi is extensively regulated by incompatibility, which determines, in the absence of any morphological differentiation, the pattern of mating among individual strains. Control of the interactions that comprise the sexual reproductive process resides in specific genetic factors, the incompatibility factors, which occur in several distinct systems in the various groups of fungi and which exert their control in two basically different ways. On the one hand, the system may play the same role as dioecy in higher organisms by restricting or preventing inbreeding among the members of the same race (*homogenic incompatibility*) and thus enhance outbreeding. On the other hand, the system may impose the opposite effect by restricting or preventing interbreeding between members of different races (*heterogenic incompatibility*) and thus promote inbreeding. In addition to these basic facts concerning the general biological significance of incompatibility, important advances have been made in recent years, especially in the investigation of the genetics of incompatibility systems. Sufficient information concerning the genetic determination of incompatibility is now available to understand many phenomena which were very mysterious in the 1920's, when H. KNIEP, of Würzburg, Germany, laid the groundwork for all subsequent study of incompatibility in the higher fungi. Furthermore, there is at present enough conceptual understanding of the physiological activity of the incompatibility-genes and of their action in morphogenetic processes to permit at least the formulation of plausible models of the operation of incompatibility systems.

On the occasion of the X. International Botanical Congress in Edinburgh, a Symposium, entitled "Incompatibility in Fungi", brought together many of the workers who are active in studies of fungal incompatibility. The Symposium was organized to provide the opportunity both for the reporting of still unpublished results as well as for the consideration of previously published work in synoptic presentations. The editors were of the opinion that it would be useful to make the materials of the Symposium available to a wider circle of workers and students and thereby to strengthen such interest as now exists in this area of research and possibly to arouse new interest. The authors of the several papers agreed to provide manuscripts of their presentations, and we took the trouble, at the time of the Symposium, to have notes taken of the discussion that followed each paper. The discussions following the individual papers below have been prepared, in collaboration with the participants, by Dr. STANLEY DICK. Responsibility for the contents of papers belongs, of course, to their individual authors, and editorial changes have been held to the minimum required for consistent terminology. The one exception to this editorial practice is the paper of Professor BURNETT, which could nòt be brought into conformity without serious alternation of meaning.

It is our pleasant obligation to thank all of those who have supported our efforts of making the Symposium available in published form by the contribution of manuscripts and items of discussion. We also thank Mrs. MURIEL WILLIAMS for secretarial assistance and Frau A. GEBAUER for both secretarial assistance and for the preparation of the Index. Finally, our thanks are also extended to the owners of Springer-Verlag, who have made this publication possible and who have fulfilled all of our wishes regarding the content and format of the following book.

K. E.   J. R. R.

Introduction

by

JOHN R. RAPER

It has now been sixty years since BLAKESLEE (1904) reported in *Rhizopus nigricans*, the common "black bread mold", the first case of obligatory cross-mating in the fungi. The required interaction between two self-sterile individuals in the process of sexual reproduction he designated *heterothallism* in distinction to *homothallism*, in which each individual had the competence to elaborate sexual organs and to complete the sexual cycle in isolation. Within a very short time, the more common species of the Mucorales were unambiguously categorized as heterothallic or homothallic, and in each heterothallic species, two and only two classes of individuals could be found. BLAKESLEE (1906) considered the two classes of individuals of heterothallic species to be differentiated in respect to sexual sign, since one would react only with the larger, so-called female gametangia of homothallic species and the other only with the smaller, male gametangia. The two classes, previously and arbitrarily designated (+) and (—), were accordingly interpreted as ♀ and ♂, respectively.

In the decades following BLAKESLEE's pioneering work, obligatory cross-mating systems have been found to occur in all of the major groups of fungi: in the Hymenomycetes independently by BENSAUDE (1918) and KNIEP (1920, 1922); in the smut fungi (Ustilaginales) by KNIEP (1919); in the biflagellate water molds (Saprolegniaceae) by COUCH (1926); in the Euascomycetes by SHEAR and DODGE (1927); in the rust fungi (Uredinales) by CRAIGIE (1927); in the yeasts (Hemiascomycetes) by WINGE and LAUSTSEN (1939a, 1939b); in the uniflagellate water molds (Blastocladiales) by HARDER and SÖRGEL (1938); and, finally, in the acellular Myxomycetes by DEE (1960) and by COLLINS (1961). The patterns of mating in these several groups differed widely in detail, but, with the exception of the heterothallic water molds, they all shared an important common feature: each provided an absolute regulation of mating competence in the total absence of any morphological differentiation (RAPER, in press).

Despite the absence of morphological differentiation between compatible strains, the early interpreters of these mating systems attributed the basic differences more often than not to sexual characters. Thus, in the 1920's, SATINA and BLAKESLEE (1928, 1929) sought through biochemical studies to confirm the femaleness and maleness of (+) and (—) strains of heterothallic mucors; KNIEP (1920, 1922) spoke of the *A* and *B* factors of certain Hymenomycetes as "sexual factors" and of their reassortment in meiosis as "sexual differentiation"; BURGEFF (1920) termed the mating pattern in the

Mucorales and in the smut, *Ustilago violacea*, as "bipolar sexuality," and BAUCH (1930) applied the comparable term, "tetrapolar sexuality" to the four-mating type system in the smut, *Ustilago longissima;* HARTMANN (1930, 1931) and VANDENDRIES (1930a, 1930b) developed a very elaborate model to rationalize the four-mating type system of the Hymenomycetes in accordance with HARTMANN's universal theory of relative sexuality; and LINDEGREN (1933, 1936) included the "sex factors" of *Neurospora crassa* among the loci examined in the first extensive tetrad analysis in fungi. Certain of these interpretations revealed a great deal of ingenuity, since, in the higher fungi, a sexual interpretation of the mating systems required a reconciliation between an indefinite number of alternate and equivalent self-sterile but cross-fertile mating types on the one hand with only two allowable sexes, male and female, on the other. By the late 1930's, according to QUINTANILHA (1939), HARTMANN had come to consider the genetic determinants of mating competence in the Hymenomycetes as "sterility factors" rather than as "sexual factors," and, so far as I have been able to determine, QUINTANILHA, in 1939, first termed the *A* and *B* factors of a tetrapolar species, *Coprinus lagopus*, the "incompatibility factors", a term that was then gaining acceptance in reference to the genetic device common in flowering plants to determine self-sterility, cross-fertility. Both the term "incompatibility factor" and the concept of incompatibility were quickly and generally adopted and have since been applied to all mating systems in fungi save those few that are obviously based on sexual dimorphism. Thus, the LINDEGRENS (1943) attributed the pattern of mating in heterothallic yeasts to an incompatibility system consisting of two alleles, *a* and *α*, at a single locus.

Incompatibility, as applied to mating systems in the fungi and in other groups of organisms, denotes the genetic determination of mating competence in the absence of or irrespective of any morphological differentiation. ESSER (1962) recognizes two basic types of incompatibility in the fungi: (a) *homogenic incompatibility*, in which sexual interaction is prevented by *like* alleles in the two presumptive mates, and (b) *heterogenic incompatibility*, in which sexual interaction is prevented by *unlike* alleles. Systems of the former type have been intensively investigated in several, divergent groups of fungi, and there is sufficient information regarding them to permit certain generalizations. Systems of the latter type, by contrast, have been specifically recognized in only a few isolated forms, and only a single system of this sort has been analysed in any detail (cf. ESSER, this Symposium).

Homogenic incompatibility probably occurs in all three classes of fungi, and it accounts for the specific patterns of obligatory cross mating that occur in a majority of heterothallic species of fungi, with the following exceptions. There is no substantial evidence of incompatibility among the aquatic Phycomycetes, an assemblage of predominantly homothallic forms that contains occasional cross-mating species as the result of the segregation of ♂ and ♀ strains. Mating competence in the heterothallic members of the Mucorales, dependent upon the segregation of two alternate alleles at a single locus (BURGEFF, 1914), reflects an incompatibility system according to certain authors (WHITEHOUSE, 1949; RAPER, 1954) and sexual differentiation according to others (ESSER and KUENEN, 1965). Among the Ascomycetes, sexual dimorphism has been reported in *Ascosphaera* (*Pericystis*) *apis* (SPILTOIR, 1955), in various members of the Laboulbeniales (THAXTER, 1896; BENJAMIN and SHANOR, 1950), and in *Stromatinia narcissi* (DRAYTON and GROVES, 1952). Otherwise, obligatory cross-mating is imposed by incompatibility in all other heterothallic members of the Ascomycetes, and, without exception, obligatory cross-mating is imposed by incompatibility systems in the various groups of the Basidiomycetes.

The distribution of incompatibility systems in cross-mating species belonging to the various subdivisions of the fungi almost exactly parallels that of the competence to form heterokaryons (RAPER, 1955). Heterokaryosis is a fungal monopoly and is unquestionably a major factor in the evolution of the higher fungi; the evolutionary significance of heterokaryosis is best exemplified by the incorporation of a specialized heterokaryon, the dikaryon, in the life cycle of the more advanced forms, transiently in the Euascomycetes and persistently in all Basidiomycetes (WHITEHOUSE, 1951). The relationship between heterokaryosis and incompatibility is quite direct in all cases, since the incompatibility system, whenever present, determines both the occurence of and the characteristics of the heterokaryons that may be formed (RAPER and ESSER, 1964).

Although most of the fungal mating systems that involve incompatibility have been known in outline for many years, it is only during the past decade or so that they have been subjected to intensive genetic study. This renewed study has revealed these incompatibility systems to have many intriguing, unsuspected features and to provide excellent experimental materials for the study of numerous basic genetic phenomena. Incompatibility is not simple in any case, and in the higher fungi it has become related, directly or indirectly, to so many aspects of the plant's biology that perhaps none of these can be adequately understood without cognizance being taken of the basic facts of the incompatibility system. The system is complex in its genetic structure, in its physiological and morphological manifestations, and, doubtlessly, in its biochemical operation. Its complexity, however, only renders it the more challenging.

This Symposium devoted to fungal incompatibility comes at a very propitious time. Recent genetic work on several incompatibility systems has shifted primary emphasis in the study of these systems from the descriptive to the analytical, and this more analytical approach has brought several of the incompatibility systems well along toward the frontiers of present-day genetic inquiry. At least certain of the incompatibility systems have been found to have unusual genetic features that are quite without precedent in microorganisms. Most genetic work with microorganisms has involved biosynthetic pathways leading to the elaboration of essential nutrilites, and it might have been expected that novel features would be found in the detailed examination of the genetic devices that have evolved for the control of mating competence. The selective pressures have doubtlessly been quite different here, and the evolutionary histories of the two basically different functions might well have led to widely divergent genetic systems. The detailed elucidation of these incompatibility systems, in and of itself, promises to be a significant contribution to modern genetics.

Possibly even greater promise is implicit in the investigation of incompatibility in relation to the many ancillary phenomena upon which it impinges. Numerous examples of the extension of the effects of incompatibility beyond the primary control of mating competence will become evident in the several papers of this Symposium. It is not my purpose here, however, to anticipate the contents of the papers to follow, but only to point out some of the prospects that may confidently be expected in the further study and understanding of fungal incompatibility.

The very striking correlation between incompatibility and heterokaryosis in all obligatorily cross-mating species of the Euascomycetes and Basidiomycetes can lead to a more precise understanding of the detailed characteristics of heterokaryons. Heterokaryons in general afford ideal systems for the study of genic actions and interactions, and the better understanding of the regulation of heterokaryosis by the incompatibility systems can appreciably augment the presently considerable experimental utility of these uniquely fungal nuclear associations.

Two phenomena that are regulary associated with heterokaryons of higher fungi are of particular interest: somatic recombination and internuclear selection. The former, the assortment and recombination of genetic characters in vegetative systems, is common throughout the Ascomycetes and Basidiomycetes, and in the latter group, under certain situations imposed by the incompatibility systems, appears to involve a variety of different mechanisms of exchange. The clarification of present uncertainties here will be both interesting and significant. Internuclear selection, the preferential pairing of specific nuclear types, under conditions that permit a range of possible pairings, appears to be at least partially determined by the incompatibility system. If this proves to be the case, it will constitute a major step in the rationalization of the course of evolution in the higher fungi.

Finally, in all incompatibility systems, a point of central interest is the means by which the incompatibility factors so precisely regulate the interactions that occur in mating. Until very recently, this regulation could be stated only in terms of the final products of the interactions, but recent work has made modest progress toward the definition of intermediate biochemical and physiological stages in the process of mating. The step by step analysis of the regulation imposed by the incompatibility factors will not be quickly achieved in any system; its achievement, however, is a worthy central goal for future studies on fungal incompatibility.

In 1920, KNIEP wrote, as a sort of one-sentence summary of the first attempt to define the pattern of mating in a tetrapolar fungus, „*Der Fall Schizophyllum ist nicht ganz klar*". In this Symposium, we might well ask whether our real understanding of incompatibility has materially changed in the forty years since KNIEP's succinct statement of frustration and challenge. If the answer is affirmative, we should be content.

References

BAUCH, R.: Über multipolare Sexualität bei *Ustilago longissima*. Arch. Protistenk. **70**, 417—466 (1930).

BENJAMIN, R.K., and L. SHANOR: The development of male and female individuals in the dioecious species *Laboulbenia formicarum*. Am. J. Botany 37, 471—476 (1950).

BENSAUDE, M.: Recherches sur le cycle évolutif et la sexualité chez les Basidiomycètes. Thèse, Neumours 156 pp. (1918).

BLAKESLEE, A.F.: Sexual reproduction in the Mucorineae. Proc. Amer. Acad. Sci. 40, 206—319 (1904).

— Zygospore germinations in the Mucorineae. Ann. Mycol. 4, 1—28 (1906).

BURGEFF, H.: Untersuchungen über Variabilität, Sexualität, und Erblichkeit bei *Phycomyces nitens*. I. Flora 107, 259—316 (1914).

— Sexualität und Parasitismus bei den Mucorineen. Ber. Deut. Botan. Ges. 38, 38—327 (1920).

COLLINS, O.R.: Heterothallism and homothallism in two Myxomycetes. Am. J. Botany 48, 674—683 (1961).

COUCH, J.N.: Heterothallism in *Dictyuchus*, a genus of the water moulds. Ann. Botany 40, 848—881 (1926).

CRAIGIE, J.H.: Discovery of the function of pycnia of the rust fungi. Nature (London) 120, 765—767 (1927).

DEE, J.: A mating type system in an acellular slime-mould. Nature (London) 185, 780—781 (1960).

DRAYTON, F.D., and J.W. GROVES: *Stromatinia narcissi*, a new sexually dimorphic Discomycete. Mycologia 44, 119—140 (1952).

ESSER, K.: Die Genetik der sexuellen Fortpflanzung bei den Pilzen. Biol. Zentr. 81, 161—172 (1962).

—, and R. KUENEN: Genetik der Pilze. Heidelberg: Springer-Verlag 1965.

HARTMANN, M.: Die Sexualität der Protisten und Thallophyten und ihre Bedeutung für eine allgemeine Sexualitäts-Theorie. Z. ind. Abst.- u. Vererb.-Lehre 54, 76—126 (1930).

— Relative Sexualität und ihre Bedeutung für eine allgemeine Sexualitäts- und eine allgemeine Befruchtungstheorie. Naturw. Rundschau 19, 8—37 (1931).

HARDER, R., and G. SÖRGEL. 1938. Über einen neuen planoisogamen Phycomyceten mit Generationswechsel und seine phylogenetische Bedeutung. Nach. Ges. Wiss. Gottingen N. F. Fachgruppen 6, Biologie 3, 119—127 (1938).

KNIEP, H.: Untersuchungen über den Antherenbrand (Ustilago violacea Pers.). Z. Botan. 11, 257—284 (1919).

— Über morphologische und physiologische Geschlechtsdifferenzierung. (Untersuchungen an Basidiomyzeten.) Verh. phys.-med. Ges. Würzburg 46, 1—18 (1920).

— Über Geschlechtsbestimmung und Reduktionsteilung. Verh. phys.-med. Ges. Würzburg 47, 1—28 (1922).

LINDEGREN, C.C.: III. Pure breed stocks and crossing over in *N. crassa*. Bull. Torrey Botan. Club 60, 133—154 (1933).

— The structure of the sex-chromosome of *Neurospora crassa*. J. Heredity 27, 250—259 (1936).

—, and G. LINDEGREN: Segregation, mutation and copulation in *Saccharomyces cerevisiae*. Ann. Mo. Botan. Garden 30, 453—468 (1943).

QUINTANILHA, A.: Étude génétique du phénomène de Buller. Bol. Soc. broter. 13, 425—486 (1939).

RAPER, J. R.: Life cycles, sexuality and sexual mechanisms in the Fungi, in Sex in Microorganisms. (Eds. WENRICH, D.H., I.F. LEWIS, and J.R. Raper.) Am. Assoc. Advance. Sci., pp. 42—81. Washington 1954.
— Heterokaryosis and sexuality in fungi. Trans. Acad. Sci. N.Y. II **17**, 627—635 (1955).
— In press. Life cycles, basic patterns of sexuality and sexual mechanisms. The Fungi (A.S. SUSSMAN, and G.C. AINSWORTH, eds.), Vol. II.
—, and K. ESSER: The Fungi. In The Cell (J. BRACHET, and A.E. MIRSKY, eds.). **6**, 139—245 (1964).
SATINA, S., and A.F. BLAKESLEE: Studies on biochemical differences between sexes in Mucors. V. Proc. Natl. Acad. Sci. (U. S.) **14**, 308—316 (1928).
— — Criteria of male and female in bread molds (Mucors). Proc. Natl. Acad. Sci. (U.S.) **15**, 735—740 (1929).
SHEAR, C.L., and B.O. DODGE: Life histories and heterothallism of the red bread mold fungi of the Monilia group. J. Agr. Res. **34**, 1019—1042 (1927).
SPILTOIR, C.F.: Life cycle of *Ascosphaera apis* (*Pericystis apis*). Am. J. Botany **42**, 501—508 (1955).
THAXTER, R.: Contribution towards a monograph of the Laboulbeniaceae. Mem. Amer. Acad. Arts Sci. **12**, 187—429 (1896).
VANDENDRIES, R.: La conduite sexuelle des Hymenomycètes interpretée par les théories de *Hartmann* convernant le bisexualité et la relativité sexuelle. Bull. Acad. Belg. Cl. Sci. **16**, 1213—1234 (1930a).
— Conduite sexuelle de Psathyrella disseminata et essais de determination des valeurs relatives des realizateurs sexuels selon *Hartmann*. Bull. Acad. Belg. Cl. Sci. **16**, 1235—1249 (1930b).
WHITEHOUSE, H.L.K.: Heterothallism and sex in fungi. Biol. Rev. **24**, 411—447 (1949).
— The significance of some sexual phenomena in the fungi. Indian Phytopath. **4**, 91—105 (1951).
WINGE, Ö., and O. LAUSTSEN: On 14 new yeast types, produced by hybridization. Compt. Rend. Trav. Lab. Carlsberg, Serie Physiologique **22**, 337—355 (1939a).
— — *Saccharomycodes Ludwigii*, a balanced heterozygote. Compt. Rend Trav. Lab. Carlsberg, Serie Physiologique **22**, 357—370 (1939b).

Heterogenic incompatibility

by

KARL ESSER

In the past, sexual incompatibility was defined from a genetic point of view, *that genetically like gametes or nuclei could not undergo karyogamy.* This basic principle was valid for all different incompatibility mechanisms found in higher plants as well as in fungi. During the last decade, however, it has become evident that this is *not the only* general principle leading to incompatibility. In the study of different geographical races of the ascomycete *Podospora anserina*, which belongs to the Sordariaceae, *it was found that the prerequisite for incompatibility can also be a heterogenic structure of the gametic nuclei.* We have suggested the terms *"homogenic incompatibility"* and *"heterogenic incompatibility"*, respectively, *to define these two restricting systems for sexual propagation on the basis of their genetic determination*[1].

Before we look at heterogenic incompatibility as it was found in *P. anserina*, we shall first consider a few peculiarities of this fungus which are necessary for the better understanding of the heterogenic incompatibility system.

Female and male sex organs (ascogonia and spermogonia respectively) are formed on the mycelia of *P. anserina*. The ascogonia are fertilized by male gametes, or spermatia, by way of a trichogyne.

This sexual reaction is controlled by *two* incompatibility systems. Within each geographical race of this fungus, one finds two mating types, + and —, which regulate sexual compatibility by the well-known bipolar mechanism

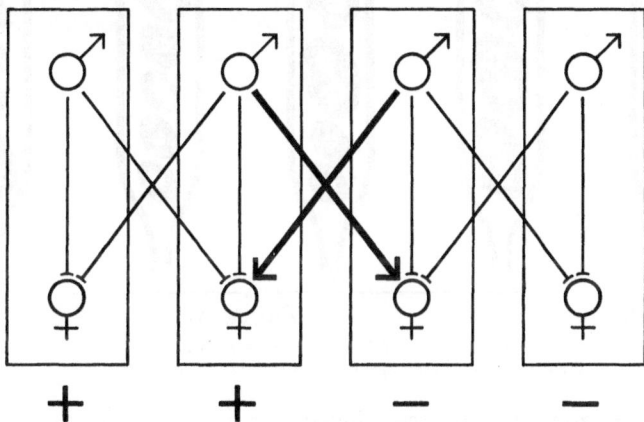

Figure 1. Scheme of bipolar homogenic incompatibility controlling the sexual reaction between +- and —- strains within each geographical race of *P. anserina*

of homogenic incompatibility. Since the formation of a zygote can only occur in the + × — combination, the line of perithecia, formed in the zone of contact between the two self-incompatible + and — strains, is a mixture of the two reciprocal crosses, +♀ × —♂ and —♀ × +♂, respectively (fig. 1 and fig. 3).

The homogenic bipolar incompatibility of *P. anserina*, as demonstrated in fig. 1, is masked in nature by a phenomenon termed pseudocompatibility, a secondary homothallism. Each ascus of *P. anserina* normally contains four binucleated spores. Due to an extremely high postreduction frequency of about 97% of the mating type alleles, most asci contain spores with +- and —-nuclei (fig. 2). The resulting heterokaryotic mycelia form perithecia and are thus self-fertile. Homokaryotic self-incompatible mycelia, however, may be obtained by two ways: 1. Prereduction of the mating type alleles (about 3%). 2. Formation of uninucleate spores, a pair of which originates in about

---

[1] All papers concerning heterogenic incompatibility in *P. anserina* are listed on page 12 and 13.

1—2% of all asci instead of single binucleate spores (fig. 2). The mycelia growing out of spores of this type are self-incompatible and behave exactly as mentioned above. In our further statements, we will only deal with this type of mycelia.

In crosses between different geographical races, one can generally observe a phenomenon which is called barrage (fig. 3). This phenomenon consists in a macroscopically visible white zone between the two mycelia. In

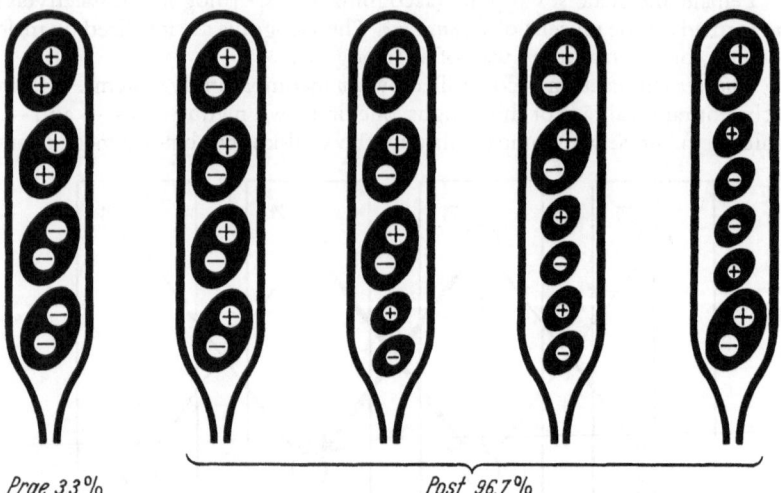

Prae 3,3%　　　　　　　　Post 96,7%

Figure 2. Scheme of different ascus types in *P. anserina*.
Prae = prereduction frequency, Post = postreduction frequency

the barrage zone, which occurs independently of the mating types, the hyphal tips of the two adjacent mycelia do not form melanin pigments. The barrage phenomenon was first discovered by Rizet (1952). In the two races analysed by Rizet and his group, the formation of a barrage is due to genic and extrachromosomal determinants. In all other cases analysed later, however, extrachromosomal differences do not occur.

In crosses between opposite mating types of different geographical races, one very often finds deviations from the reciprocal compatibility between +- and —-mating types.

1. A non-reciprocal incompatibility, which has been called semi-incompatibility: i.e. only one of the two reciprocal crosses is compatible, the other is incompatible (see fig. 3 upper and lower half).

2. A reciprocal incompatibility: Between +- and —-strains of different races no perithecia are formed (fig. 3, right half).

A very essential tool in identifying semi-incompatibility is the formation of a barrage.

As we can further deduce from figure 3, four loci (*a b c v*), each with two alleles, are responsible for semi-incompatibility and for total incompatibility. The four genes are not linked. Two genes at a time are responsible for semi-incompatibility, which occurs only in the crosses *ab* × *a1 b1* and *cv* × *c1 v1* respectively. Total incompatibility comes about by overlapping of the two semi-incompatibility mechanisms (*abc1v1* × *a1b1cv*). Therefore, the basis of total incompatibility is semi-incompatibility. Since incompati-

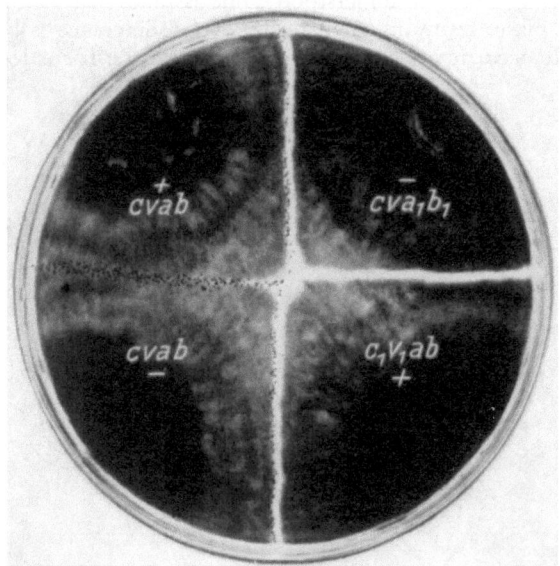

Figure 3. The genetic basis of heterogenic incompatibility in P. anserina. A petri dish has been inoculated with four different mycelia. The genotypes of the four strains are noted. For further explanations see text (from ESSER, 1956)

bility occurs only when the partners differ in the *ab* or *cv* genes, we should call this kind of incompatibility *heterogenic*.

To sum up this point, we may say that *the sexual behavior of P. anserina is determined by both homogenic incompatibility and heterogenic incompatibility. The first works within each race and between races, while the latter acts only between different races and in addition to the basic homogenic +/— mechanism.*

Heterogenic incompatibility does not only operate in the sexual phase but also in the vegetative. By specific combinations of the incompatibility genes, the vitality of nuclei is diminished in homokaryons as well as in heterokaryons. Two cases might be considered:

1. Homokaryotic mycelia having the genotypes *a1b* or *c1v*, which occur as recombinant types in semi-incompatible crosses, grow very sparsely. In their hyphae, the nuclei degenerate very rapidly.

2. The genes *a1* and *b* or *c1* and *v* are incompatible not only when they are in the same nucleus but also when in different nuclei in heterokaryons. Such heterokaryons, e. g. *ab* + *a1b1*, exhibit normal growth behavior, but the nuclei carrying the gene *a1* exhibit a very low division rate and are therefore very soon eleminated. The same holds true for the *c1 v1* nuclei in the *cv* + *c1 v1* heterokaryon.

In both cases, there exists an antagonism between *b* and *a1* (or *v* and *c1*), in which the normal action of nuclei carrying *a1* (or *c1*) is suppressed.

These observations allow us to give an explanation for the asymmetric fertilization in semi-incompatible crosses as shown in fig. 4 for the *ab/a1 b1* case.

This phenomen can be understood if one assumes that, due to the *b—a1* interaction, plasmogamy initiated by the +/— difference is blocked, but this can only occur if *b* is present in the male gamete. If *b* is located in the

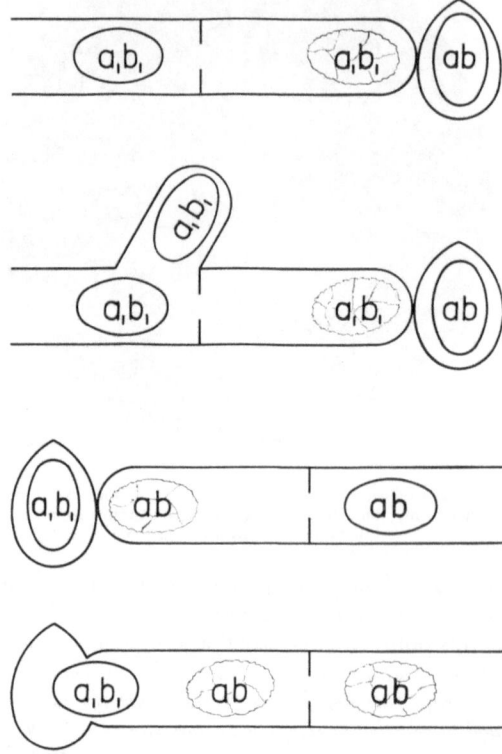

Figure 4. Schematic representation of a model to explain the asymmetric behavior in fertilization due to heterogenic incompatibility between trichogyne tips and spermatia. For explanations see text.

⊕ = degenerating nucleus   ◯ = normal nucleus

female partner, its action cannot take place, because in the trichogyne the nuclei normally start to degenerate as soon as the trichogyne tip comes in contact with a spermatium. In this case fertilization takes place.

The *biochemical actions* of *b* and *v*, respectively, were characterized by means of immunological experiments. We were able to show that, in mixed, partially heterokaryotic cultures, the protein spectrum differs from that found in extracts from homokaryotic cultures.

BERNET (1963) has recently analysed heterogenic incompatibility between two other races of *P. anserina*. In principle, he confirmed the results described above. He was able to identify three pairs of alleles (*C/c, D/d, E/e*), of which two pairs at a time led to semi-incompatibility in heterogenic combinations (i. e. $Ce \times cE$ and $Cd \times cD$ resp.). He also found the same irregularities in the vegetative phase mentioned above. The only difference between his findings and ours is due to the fact that the action of *gene D* (which can be compared to our genes *b* or *v*) is dependent on temperature. It is only manifested at 20° C, but not at 32° C. It would be very interesting to know, whether the incompatibility-genes found by BERNET are allelic with our genes, or if there exist in *P. anserina* still other loci that control the same phenomenon.

Heterogenic incompatibility is not necessarily related to homogenic incompatibility. In the self-compatible fungi, *Sordaria fimicola* (OLIVE, 1956) and *S. macrospora* (ESSER, unpublished work), heterogenic incompatibility also occurs between different races. For example, OLIVE was able to show, in interrace crosses of *S. fimicola*, that formation of perithecia did not always occur. While races *A1* and *C1* could not be crossed, both races were fertile with a third one (*C4*). One may therefore conclude that the 'sterility' between *A1* and *C1* is not due to cross-sterility (chromosomal differences etc.) but to real incompatibility. Since monosporous mycelia of each race are self-compatible, one can further suggest that the inter-race incompatibility may be due to heterogeneity.

In a search of the relevant literature, we found especially in older papers some unexplained experiments concerning 'cross-sterility' in fungi, higher plants, and even animals, which may easily be explained by heterogenic incompatibility. (For references, see ESSER and KUENEN, 1965). Unfortunately, in all these examples, no detailed genetic analyses have been made which would permit the formulation of models to explain the gene-action of heterogenic incompatibility.

It therefore seems that heterogenic incompatibility is of widespread occurrence among living beings. Further detailed genetic analysis of more examples might lead to a more comprehensive knowledge about this phenomenon.

Let us just devote a few minutes for a discussion of the general meaning of heterogenic and homogenic incompatibility for evolution.

The efficiency of sexual propagation as a means for recombination of genetic material increases as fertilization between genetically like nuclei is diminished. This restriction of inbreeding is favoured by dioecy. The same effect can also be obtained in monoecious forms by homogenic incompatibility, because, in all known mechanisms of this system, genetically like nuclei are incompatible. Homogenic incompatibility therefore, has to be considered as a system which diminishes inbreeding and enhances outbreeding.

Heterogenic incompatibility, however, has just the contrary effect in evolution. Since this system is based on the heterogeneity of gametic nuclei, it does not occur within a single race. Heterogenic incompatibility enhances inbreeding and diminishes outbreeding. It has to be considered as an isolating mechanism.

In conclusion, one can say: Within a species, the sexual behavior of which is determined by *dioecy or homogenic incompatibility*, there are *very poor possibilities of inbreeding.* Therefore the genetic material is constantly recombined. Mutations which occur spontaneously will be rapidly distributed within that *species*, which *participates as a whole in evolution.* On the contrary, this does not hold true in species whose races show *heterogenic incompatibility.* Here, mutations are only very slightly transferred from one race to the other. In this case, *not the species, but the race seems to be the smallest unit of evolution.*

References

Papers on heterogenic incompatibility of *Podospora anserina:*

RIZET, G., et K. ESSER: Sur des phénomènes d'incompatibilité entre souches d'origines différentes chez *Podospora anserina.* Compt. Rend. Acad. Sci. (Paris) **237**, 760—761 (1953).

ESSER, K.: Sur le déterminisme génétique d'un nouveau type d'incompatibilité chez *Podospora.* Compt. Rend. Acad. Sci. (Paris) **238**, 1731—1733 (1954).

— Genetische Analyse eines neuen Incompatibilitätstypes bei dem Ascomyceten *Podospora anserina.* Compt. Rend. 8. Congr. Intern. Bot. (Paris) Sect. **10**, 72—77 (1954).

— Genetische Untersuchungen an *Podospora anserina.* Ber. Deut. Botan. Ges. **68**, 143—144 (1955).

— Die Incompatibilitätsbeziehungen zwischen geographischen Rassen von *Podospora anserina* (CES.) REHM. I. Genetische Analyse der Semi-Incompatibilität. Z. indukt. Abstamm.- u. Vererb.-L. **87**, 595—624 (1956).

— The significance of semi-incompatibility in the evolution of geographic races in *Podospora anserina.* Proc. X. Intern. Congr. Genetics, Vol. II, 76—77 (1958).

— Die Incompatibilitätsbeziehungen zwischen geographischen Rassen von *Podospora anserina* (CES.) REHM. II. Wirkungsweise der Semi-Incompatibilitäts-Gene. Z. Vererbungsl. **90**, 29—52 (1959).

— Die Incompatibilitätsbeziehungen zwischen geographischen Rassen von *Podospora anserina* (CES.) REHM. III. Untersuchungen zur Genphysiologie der Barragebildung und der Semi-Incompatibilität. Z. Vererbungsl. **90**, 445—456 (1959).

BERNET, J., K. ESSER, D. MARCOU, et J. SCHECROUN: Sur la structure génétique de l'espèce *Podospora anserina* et sur l'interèt de cette structure pour certain recherches de génétique. Compt. Rend. Acad. Sci. (Paris) **250**, 2053—2055 (1960).

ESSER, K.: Incompatibilität bei Pilzen. Ber. Deut. Botan. Ges. **74**, 324—325 (1961).

— Die Genetik der sexuellen Fortpflanzung bei den Pilzen. Biol. Zbl. **81**, 161—172 (1962) (Review).

BERNET, J.: Action de la température sur les modifications de l'incompatibilité cytoplasmique et les modalités de la compatibilité sexuelle entre certaines souches de *Podospora anserina.* Ann. Sci. Nat. Botan. (Paris) 12, Serie **IV**, 205—224 (1963).

Further references:

ESSER, K., u. R. KUENEN: Genetik der Pilze. Heidelberg: Springer-Verlag 1965.

OLIVE, L. S.: Genetics of *Sordaria fimicola.* I. Ascospore color mutants. Am. J. Botany **43**, 97—107 (1956).

Rizet, G.: Les phénomènes de barrage chez *Podospora anserina*. I. Analyse génétique des barrages entre souches *S* et *s*. Rev. Cytol. Biol. Vegetales **13**, 51—92 (1952).

Discussion

Day: Is there any interaction between the two systems of homogenic and heterogenic incompatibility?

Esser: No. Each exists independently of the other.

Burnett: What happens when you combine the *a1b* and *c1v* system with Bernet's system?

Esser: We don't know if our genes are identical with the genes detected by Bernet. It would be very interesting, however, to test the possibility of whether, in distinction to the homogenic system, multiple allelic genes are involved in the heterogenic system.

Snider: In the heterokaryon where one type of nucleus is soon eliminated, do you interpret this process as a specific block against nuclear division of the minority-type nucleus?

Esser: Yes.

Day: What happens if you cross a fully fertile heterokaryon with a homokaryon which has heterogenic semi-incompatibility factors? Is there any interaction in the fully compatible heterokaryon which might override the heterogenic incompatibility?

Esser: No. Both nuclear components of the heterokaryon act independently with the homokaryon.

Homogenic incompatibility

Incompatibility in yeasts

by

Majeed Ahmad

Sexuality in yeast was discovered in 1900 by Hoffmeister. Satava (1918, 1934); Winge (1935, 1944); and Winge and Laustsen (1937, 1938, 1939) found a regular alternation of haploid and diploid phases in the life-cycle of certain yeasts. Winge (1935) maintained that the yeasts were homothallic, whereas Lindegren and Lindegren (1943 a, b, and c) claimed that they were heterothallic.

Ahmad (1953) showed that some strains of Saccharomyces were heterothallic (*S. cerevisiae*, Danish Baking Yeast, *S. carlsbergensis*) and others homothallic (*S. cerevisiae*, Brewer's Yeast). The heterothallic strains gave a false appearance of homothallism as a result of mutation and the fusion between altered and unchanged nuclei.

A. Genetics of incompatibility factors

Incompatibility in yeasts is controlled by one locus with two allelomorphs. These alleles have been represented differently by different authors. Lindegren and Lindegren (1943a, b, c), and Takahashi (1958 and 1961)

have named them a and a; AHMAD (1948, and 1953) first represented them
as — and + and later as a and β (AHMAD and KHAN, 1955a). LEVI (1956)
represented them as — and +.

In *Schizosaccharomyces pombe*, LEUPOLD (1950) found that the mating-
type locus had a series of heterothallic and homothallic alleles, h^N, h^R, h—,
h^{90}, h^{40} and a sterile allele. He found evidence of intragenic recombination
and concluded that the mating type locus in this yeast consisted of two
parts with a distance of one cross over unit between them (LEUPOLD, 1958).
Both mating-type loci, A and B, in *Schizophyllum commune* and *Coprinus
lagopus* have been found to consist of 2 loci (RAPER, 1959, RAPER et al. 1960,
PARAG, 1962). It seems likely that in *Schizosaccharomyces* the two parts of the
mating-type locus are not 2 separate genes but two subunits of a single gene.

B. Monogenic secondary homothallism or the D-gene system

WINGE and ROBERTS (1949) established that single spore cultures of
Saccharomyces chevalieri possessed a gene, D, for selfdiploidisation, while the
heterothallic strain of *S. cerevisiae* carried the corresponding allele, d. They
also showed that D, d segregated independently of the two mating-type
alleles A, a.

TAKAHASHI, SAITO, and IKEDA (1958); TAKAHASHI (1958, 1961 and
1963); and TAKAHASHI and IKEDA (1959) showed that some of the strains
classified as homothallic possessed a mating reaction, which was not revealed by testers (mass mating technique) but was detectable only by special
means, the "minimal plate mating technique".

This technique revealed that the homothallic strains carrying the gene
D gave a cryptic mating-type reaction with both the testers. When the D
strain was mated with A strain, it acted as a, and when the D strain was
mated with a it acted as A. TAKAHASHI (1963) therefore concluded that
gene D converted unisexual heterothallic strains into bisexual homothallic
strains. When a single D spore was mated with an A spore, the hybrid
segregated not only for D and A but also for a. Similarly, crosses of
D with a segregated for A. The D cultures therefore carried both A and
a mating-type alleles, and D must act as a mutator of the mating-type locus.
(TAKAHASHI and IKEDA, 1959).

C. Multigenic secondary homothallism or the HM-gene system

Through use of the "minimal plate mating technique", TAKAHASHI and
IKEDA detected that some homothallic strains were unisexual diploids
possessing either an A or an a concealed mating-type reaction. They claimed
that this type of homothallism was under the control of three loci, HM_1,
HM_2, and HM_3, which were linked to the mating-type locus. If only one
of the three dominant genes, HM_1, HM_2 or HM_3, was present, the culture
gave a mass mating reaction, but if two were present, e.g., HM_1 and HM_2
or HM_2 and HM_3, the culture gave a concealed mating type reaction.
TAKAHASHI (1958) has proposed that the concealing of the mass mating
reaction is the result of the complementary action of two genes. Crosses of

$HM_1 \, hm_2 \, hm_3 \, \chi \, hm_1 \, HM_2 \, hm_3$ segregated in the ratio of one homothallic: three heterothallic single spore cultures. If this gene system consists of three pairs of genes, however one should be able to recover 8 genotypes in all, but TAKAHASHI obtained only 6.

It seems that the Hm genes act as a set of polygenic suppressors in the mating reaction. One dominant gene brings about apparently no suppression, but two together have a strong complementary action to leave only a trace of the mating potency, which is expressed as a concealed mating-type. The cells at the same time acquire the capacity to diploidize and form spores. These spores are of one mating-type. The HM genes promote self diploidization and not mutation to the opposite mating-type allele. Therefore the action of the HM genes is different from the D gene, and TAKAHASHI has shown that the D gene segregated independently of the HM genes.

D. Mutation of the mating-type alleles

The mating-type alleles in yeast are subject to mutation, (AHMAD, 1948, 1952, 1953; LINDEGREN and LINDEGREN, 1943c and 1944; LEUPOLD, 1950; OESER, 1962), and these mutations usually occur in the post meiotic phases. We have occasionally observed segregations of the mating-type alleles with a ratio of $1A : 3a$ and $3A : 1a$ (tab. 1). These rare departures from the $2A : 2a$ segregation may be due to the mutation of the mating-type alleles during meiosis. They could also result from a mitosis following meiosis and random abortion of four out of the eight spores in an ascus. I have only once observed more than four spores in an ascus in the yeast strains with which I am working.

Table 1. *Anomalous segregations of mating-type alleles in tropical strains of Saccharomyces carlsbergensis*

Number of spores germinated per ascus	Type of segregation				Number of asci showing segregation
4	$2A$	:	$2a$		33
	$3A$	:	$1a$		2
	$1A$	:	$3a$		2
	$1A$	:	$2a$	: $1N$	1
	$1A$	:	$2a$		65
3	$2A$	:	$1a$		59
	$1A$	:	$1a$	: $1N$	1
	$0A$	:	$3a$		2

In the different strains of yeast, there seems to be differences in the rates of mutation of their mating type alleles (tab. 2). The tropical strains of *Saccharomyces carlsbergensis* appear to have a much lower mutation rate than the European heterothallic yeasts. The mutation rate in the two mating type alleles is not the same in the Danish Baking Yeast. a mutates to A at about twice the rate that A mutates to a (tab. 3).

Table 2. *Lower mutation rate of mating-type alleles in the tropical strains of Saccharomyces carlsbergensis*

Region	Strain *Saccharomyces* species	Total number of single spore cultures analysed	Total number of non-mating spores	Percentage of non-mating spores
East Pakistan	*S. carlsbergensis*	1432	6	0.42
Europe	*S. cerevisiae* (Danish Baking Yeast)	182	56	3.7
	S. cerevisiae strain 237	12	4	33.3
	S. carlsbergensis	4	1	25

Table 3. *Mutations in a and A mating-type alleles of Danish Baking Yeast*

Mating-type	Total	Heterothallic single spore cultures analysed		
		Sporulating (Mutant)	Non-sporulating (non-mutant)	Percentage of mutants
A	96	25	71	26
a	61	34	27	56

Loss of mating reaction by yeasts during vegetative multiplication is also assigned to mutation of the mating-type alleles. Aneuploidy (TAKA-HASHI and IKEDA, 1959) and triploidy (WINDISCH, 1962) have also been claimed to be the cause of sterility of cultures.

E. Modifying genes

It appears that the yeasts carry modifiers for the incompatibility factors. In an analysis of the mating systems in yeast, AHMAD (1948) observed

Table 4. *Differences in the mating reaction of single spore cultures of Saccharomyces cerevisiae. (Danish Baking Yeast)*

MATING		REACTION			
STRONG		WEAK		NONE	
Sporulating	Non-sporulating	Sporulating	Non-sporulating	Sporulating	Non-sporulating
16	92	13	6	47	9

non-sporulating cultures with strong reaction, weak reaction, and no reaction (tab. 4). WINGE and ROBERTS (1949) also found evidence for the presence of modifiers when they analysed a cross between the homothallic strain of *Saccharomyces chevalieri* and the heterothallic strain of *S. cerevisiae*. Some of the single spore cultures from this cross showed a weaker mating reaction than the testers.

F. Control of sexual reproduction in Saccharomyces

The mating reaction in *Saccharomyces* involves several steps, which commence with the stimulation of the cells and culminate in the fusion of the nuclei. The haploid yeast cells may be sterile, homothallic, or hetero-thallic of mating-type A or of mating-type a. The two mating-type alleles A and a can mutate from one form to the other. It seems likely to the author that the mating reaction in Saccharomyces is controlled by an operon, two complementary structural genes, one regulator, and one inhibitory locus with three alleles which probably operate as the mating-type locus.

G. Mating systems in yeast

Data on the mating systems of about 133 strains are now available (tab. 5). These strains belong to 11 genera and 58 species. Of the 11 genera, 2 include only homothallic species, 5, both homothallic and heterothallic species, and 4, only heterothallic species. Some of the species include both homothallic and heterothallic strains (tab. 5). The yeasts thus differ in their mating system not only in different species of a genus but also in different strains of a species. An examination of the available data reveals the homo-thallic and the obligate heterothallic genera have very few species, while the partial heterothallic genera have a large number of species.

In yeasts, as in other fungi, undifferentiated sexuality or primary homo-thallism may be taken to be primitive (WHITEHOUSE, 1949, RAPER, 1959). From some of the primary homothallic forms, heterothallic types with two complementary strains may have come into being. Heterothallism must have been favoured by natural selection, because of the advantages occurring from recombination among a much larger number of genes. At times, however, heterothallic strains ran the risk of being deprived of sexual reproduction, so some of these evolved mechanisms whereby sexual repro-duction was certain to take place, even if a compatible mate was missing. Three of these mechanisms are mutations of the mating-type alleles, the D-gene system, and the HM-gene system.

H. Incompatibility factors and taxonomy of yeast

WICKERHAM and BURTON (1952) found that in nature some species existed predominately as haploid mating-types. This observation suggested that individual mating-types may have been classified in the nonsporoge-

Table 5. *The mating system of different yeasts*

Genera No.	Genera Name	Species No.	Species Name	Homothallic	Heterothallic	References
1.	Hanseniaspora	1.	valbyensis	1	0	Miller and Phaff, 1958
	Hanseniaspora	2.	uvarum	1	0	Miller and Phaff, 1958
2.	Kluyveromyces	3.	africanus	1	0	Roberts, 1960
3.	Saccharomyces	4.	occidentalis	1	0	Hjort, 1956
	Saccharomyces	5.	marxianus Hansen	1	0	Hjort, 1956
	Saccharomyces	6.	paradoxus Batschinskaia	1	0	Hjort, 1956
	Saccharomyces	7.	italicus	1	0	Ditlevsen, 1944
	Saccharomyces	8.	pini	1	0	Slodki, Wickerham, Cadmus, 1961
	Saccharomyces	9.	florentinus (Castelli)	2	0	Windisch, 1962
	Saccharomyces	10.	willianus Saccardo	3	0	Windisch, 1962
	Saccharomyces	11.	oviformis	5	0	Windisch, 1962
	Saccharomyces	12.	bayanus Saccardo	6	0	Windisch, 1962
	Saccharomyces	13.	monacensis	1	0	Takahashi, 1961
	Saccharomyces	14.	batatae	1	0	Takahashi, 1961
	Saccharomyces	15.	ellipsoideus (Hansen) forma Johannisberg II	1	0	Winge, 1935
	Saccharomyces	16.	validus Hansen	1	0	Winge, 1935
	Saccharomyces	17.	marchalianus Kufferath	1	0	Winge, 1935
	Saccharomyces	18.	pastorianus	20	0	Windisch, 1962
	Saccharomyces	19.	cerevisiae	1	0	Takahashi, 1958
	Saccharomyces		cerevisiae	1	1	Lindegren and Lindegren, 1946
	Saccharomyces		cerevisiae	8	4	Ahmad, 1953
	Saccharomyces		cerevisiae	0	1	Takahashi, 1961
	Saccharomyces		cerevisiae	0	1	Ahmad, 1948
	Saccharomyces		cerevisiae var. ellipsoideus (CX)	0	1	Winge & Roberts, 1949
	Saccharomyces		cerevisiae var. ellipsoideus (CX)	1	1	Ahmad and Khan, 1955b
	Saccharomyces	20.	carlsbergensis	4 (polyn)	0	Windisch, 1962
	Saccharomyces		carlsbergensis	0	1 (3n)	Oeser, 1962
	Saccharomyces		carlsbergensis	0	1	Ahmad and Khan, 1955b
	Saccharomyces		carlsbergensis	0	4 (2n)	Ahmad, 1948
	Saccharomyces		carlsbergensis	0	2	Oeser, 1962
	Saccharomyces	21.	chevalieri	2	2	Takahashi, 1961
	Saccharomyces		chevalieri	1	0	Takahashi, 1961
	Saccharomyces		chevalieri	1	0	Winge and Roberts, 1949
	Saccharomyces		chevalieri	1	0	Ahmad et al, 1959
	Saccharomyces		chevalieri	2	1	Ahmad et al, 1954

Genus	No.	Species	(94)	(39)	Reference
Saccharomyces	22.	rouxii	1	0	Hjort, 1956
Saccharomyces	22.	rouxii	0	1	Wickerham & Burton, 1960
Saccharomyces	23.	uvarum Beijerinck	3	1	Windisch, 1962
Saccharomyces	24.	species form 237	0	1	Ahmad, 1953
Saccharomyces	25.	kluyveri	0	1	Phaff, Miller and Shifrine, 1956
Saccharomyces	26.	diastaticus	0	1	Wickerham, 1958
Saccharomyces	27.	logos (van Laer & Denamur)	1	0	Windisch, 1962
4. Schizosaccharomyces	28a.	pombe	2	0	Ahmad et al., 1954
	28b.	pombe str.liquefaciens	1	0	Leupold, 1950
5. Zygosaccharomyces	29.	ashbyi	1	0	Wickerham & Burton, 1956a
Zygosaccharomyces	30.	dobzhanskii	1	1	Wickerham & Burton, 1956b
Zygosaccharomyces	31.	priorianus	1	0	Winge & Laustsen, 1939a
Zygosaccharomyces	32.	lactis	1	0	Wickerham & Burton, 1952
6. Hansenula	33.	canadensis	1	0	Wickerham & Burton, 1952, 1954 and 1962
Hansenula	34.	minuta	1	0	Wickerham & Burton, 1952, 1954 and 1962
Hansenula	35.	capsulata	1	0	Wickerham & Burton, 1952, 1954 and 1962
Hansenula	36.	silvicola	1	0	Wickerham & Burton, 1952, 1954 and 1962
Hansenula	37.	angusta	1	0	Wickerham & Burton, 1952, 1954 and 1962
Hansenula	38.	californica	1	0	Wickerham & Burton, 1952, 1954 and 1962
Hansenula	39.	morakii	1	0	Wickerham & Burton, 1952, 1954 and 1962
Hansenula	40.	suaveolens	1	0	Wickerham & Burton, 1952, 1954 and 1962
Hansenula	41.	saturnus	1	0	Wickerham & Burton, 1952, 1954 and 1962
Hansenula	42.	beijerinckii	1	1	Wickerham & Burton, 1952, 1954 and 1962
Hansenula	43.	wingei	0	1	Wickerham & Burton, 1952, 1954 and 1962
Hansenula	44.	beckii	0	1	Wickerham & Burton, 1952, 1954 and 1962
Hansenula	45.	holstii	0	1	Wickerham & Burton, 1952, 1954 and 1962
Hansenula	46.	subpelliculosa	0	1	Wickerham & Burton, 1952, 1954 and 1962
Hansenula	47.	anomala	0	1	Wickerham & Burton, 1952, 1954 and 1962
Hansenula	48.	ciferrii	0	1	Wickerham & Burton, 1952, 1954 and 1962
7. Dekkeromyces	49.	species	1	0	Wickerham & Burton, 1956a
Dekkeromyces	50.	species	1	0	Wickerham & Burton, 1956a
Dekkeromyces	51.	lactis	1	1	Wickerham & Burton, 1956a
Dekkeromyces	52.	fragilis	0	0	Wickerham & Burton, 1956a
8. Dematium	53.	chodati	0	1	Wickerham & Burton, 1952
9. Citeromyces	54.	matritensis	0	1 (2n)	Santa Maria, 1957
10. Endomycopsis	55.	guilliermondii	0	1	Wickerham & Burton, 1954
Endomycopsis	56.	ohmeri	0	1	Wickerham & Burton, 1954
Endomycopsis	57.	scolyti	0	1	Phaff & Yoneyama, 1961
11. Saccharomycodes	58.	ludwigii	0	1	Burnett, 1956, Hjort, 1954, and Winge & Laustsen, [1939b]
Total		58	94	39	

2*

nous genera. When investigations were made involving named strains, this was found to be true. Thus *Candida guilliermondii var. membranaefaciens* and *C. chalmersi* are imperfect forms of *Endomycopsis ohmeri*, and *Hansenula schneggii* and *Candida pelliculosa* are the two mating-types of *H. anomala* (WICKERHAM and BURTON, 1954).

I. Summary

Incompatibility is controlled by a single locus with two alleles in *Saccharomyces* and a series of alleles in *Schizosaccharomyces*. The mating-type alleles are subject to mutation, and different strains of yeast differ in the mutation rate of the mating-type alleles. In the Danish Baking Yeast, mating-type *a* mutates at about twice the rate of *A*.

Besides the mating-type locus, a gene *D* and a set of *HM* genes take part in the reproductive processes of some yeasts. The gene *D* brings about mutation to the opposite mating-type allele and leads to diploidisation. The *HM* genes produce unisexual diploid single spore cultures with a greatly reduced mating potency.

The mating systems of 133 strains of yeast have been summarised, and some of the evolutionary trends have been discussed.

References

AHMAD, M.: Genetics of Yeast. Ph. D. Thesis, University of Cambridge (1948).

— Single spore cultures of heterothallic *Saccharomyces cerevisiae* which mate with both the tester strains. Nature **170**, 546—547 (1952).

— The mating system in *Saccharomyces*. Ann. Botany **17**, 329—342 (1953).

—, A.R. CHOWDHURY, and K.U. AHMAD: Studies on Toddy Yeast. Mycologia **45**, 708—720 (1954).

—, and Md. A. KHAN: Genetics of Yeast. Pakistan. J. Sci. Res. **7**, 134—140 (1955a)

—, and A.A. KHAN: Studies on Baker's Yeast of East Pakistan. Mycologia **47**, 3, 329—338 (1955b).

—, Md. A. KHAN, and Md. EHTISHAMUDDIN: Studies on Toddy Yeast II. Scientist Pakistan **3**, 9—19 (1959).

BURNETT, J.H.: The Mating systems of fungi I. New Phytologist **55**, 50—90 (1956).

DITLEVSEN, E.: A case of simple segregation in *Saccharomyces italicus*. Compt. Rend. Trav. Lab. Carlsberg, Serie Physiologique **24**, 31—37 (1944).

HJORT, A.: Notes on *Saccharomyces rouxii* BOURTROUX and other yeasts with special regard to their life cycles. Compt. Rend. Trav. Lab. Carlsberg, Serie Physiologique **26**, 161—179 (1956).

— Some studies on the genus *Saccharomycodes* HANSEN. Compt. Rend. Trav. Lab. Carlsberg, Serie Physiologique **25**, 259—284 (1954).

HOFFMEISTER, C.: Zum Nachweis des Zellkerns bei *Saccharomyces*. Sitzber. Deut. nature. Med. Ver. Lotos. No. 5 (1900).

LINDEGREN, C.C.: The genetics of *Neurospora* II. Segregation of sex factors in asci of *N. crassa*, *N. sitophila*, and *N. tetrasperma*. Bull. Torrey Botan. Club **59**, 119—138 (1932).

LINDEGREN, C.C., and G. LINDEGREN: Legitimate and illegitimate mating in *Saccharomyces cerevisiae*. Genetics **28**, 81 (1943a).

— — Selecting, inbreeding, recombining and hybridising commercial yeasts. J. Bacteriol. **16**, 405—419 (1943b).

— — Segregation, mutation and copulation in *S. cerevisiae*. Ann. Mo. Botan. Garden **30**, 453—464 (1943c).

— — Instability of mating-type alleles in *Saccharomyces*. Ann. Mo. Botan. Garden **31**, 203—216 (1944).

— — The Cytogene Theory. Cold Spring Harbor Symp. Quant. Biol. **11**, 115—129 (1946).

LEUPOLD, U.: Die Vererbung von Homothallie und Heterothallie bei *Schizosaccharomyces pombe*. Compt. Rend. Trav. Lab. Carlsberg, Serie Physiologique **24**, 381—480 (1950).

— Studies on recombination in *Schizosaccharomyces pombe*. Cold Spring Harbor Symp. Quant. Biol. **23**, 161—170 (1958).

LEVI, J.D.: Mating reaction in Yeast. Nature **177**, 753—754 (1956).

MILLER, M.W., and H.J. PHAFF: A comparative study of the apiculate Yeasts. Mycopathol. Mycol. Appl. **10**, 113—141 (1958).

OESER, H.: Genetische Untersuchungen über das Paarungstypverhalten bei *Saccharomyces* und die Maltose - Gene einiger untergäriger Bierhefen. Arch. Mikrobiol. **44**, 47—74 (1962).

PARAG, Y.: Mutations in the B incompatibility factor of *Schizophyllum commune*. Proc. Nat. Acad. Sci. (U.S.) **48**, 743—750 (1962).

PHAFF, H.J., M.W. MILLER, and M. SHIFRINE: Antonie van Leeuwenhoek J. Microbiol. Serol. **22**, 145 (1956).

—, and M. YONEYAMA: Endomycopsis scolyti, a new heterothallic species of yeast. Antonie van Leeuwenhoek **27**, 196—202 (1961).

RAPER, J.R.: Sexual versatility and evolutionary processes in fungi. Mycologia **51**, 107—124 (1959).

—, M.G. BAXTER, and A.H. ELLINGBOE: The genetic structure of the incompatibility factors of *Schizophyllum commune*: The A-Factor. Proc. Nat. Acad. Sci. (U.S.) **46**, 833—842 (1960).

ROBERTS, C.: The life cycle of *Kluyveromyces africanus*. Compt. Rend. Trav. Lab. Carlsberg, Serie Physiologique **31**, 325—341 (1960).

SANTA MARIA, J.: Un nuevo genero de levaduras: *Clteromyces*. Inst. nacl. invest. agron. (Madrid) Sec. Bioquim **17**, 269—276 (1957).

SATAVA, J.: O redukovanych formach kvasinek. V. Prague, **48** (1918). (Cited from Winge: not available).

— Les formes sexuelles et asexuelles des levures et leur pouvoir fermentatif. III. Congr. Intern. Tech. Chim. Ind. Agr. Paris (1934). (Cited from Winge: not available).

SLODKI, M.E., L.J. WICKERHAM, and M.C. CADMUS: Phylogeny of phosphomannan producing yeasts II. Phosphomannan properties and taxonomic relationships. J. Bacteriol. **82**, 269—274 (1961).

TAKAHASHI, T.: Complementary genes controlling homothallism in *Saccharomyces*. Genetics **43**, 705—714 (1958).
— Sexuality and its evolution in *Saccharomyces*. Seiken Ziho **12**, 11—20 (1961).
— Genetic consideration on sexuality of *Saccharomyces*. Abstract the 13th Meeting of the Seminar of Yeast Studies, October, **16**, (1963).
—, H. SAITO, and Y. IKEDA: Heterothallic behaviour of a homothallic strain in *Saccharomyces* Yeast. Genetics **43**, 249—260 (1958).
—, and Y. IKEDA: Bisexual mating reaction in *Saccharomyces chevalieri*. Genetics **44**, 375—376 (1959).
WICKERHAM, L. J.: Sexual agglutination of heterothallic Yeasts in diverse taxonomic areas. Science **128**, 1504—1505 (1958).
—, and K. A. BURTON: Occurrence of yeast mating-type in nature. J. Bacteriol. **63**, 449—451 (1952).
— — A clarification of the relationship of *Candida Guilliermondii* to other yeasts by a study of their mating-types. J. Bacteriol. **68**, 594—597 (1954).
— — Hybridisation studies involving *Saccharomyces lactis* and *Zygosaccharomyces ashbyi*. J. Bacteriol. **71**, 290—295 (1956a).
— — Hybridisation studies involving *Saccharomyces fragilis* and *Zygosaccharomyces dobzhanskii*. J. Bacteriol. **71**, 296—302 (1956b).
— — Heterothallism in *Saccharomyces rouxii*. J. Bacteriol. **80**, 492—495 (1960).
— — Phylogeny and biochemistry of the genus *Hansenula*. Bacteriol. Rev. **26**, 382—397 (1962).
WINDISCH, S.: Microbiological problems in brewing. The Brewers Dig. February, 1962, 47—51 (1962).
WINGE, Ö.: On haplophase and diplophase in some Saccharomycetes. Compt. Rend. Trav. Lab. Carlsberg, Serie Physiologique **21**, 4, 77—111 (1935).
— On Segregation and Mutation in Yeast. Ibid. **29**, 79—95 (1944).
—, and LAUSTSEN, O.: On two types of germination and on genetic segregations in *Saccharomyces*, demonstrated through single spore cultures. Ibid. **22**, 99—116 (1937).
— — Artificial species hybridisation in yeast. Ibid. **22**, 235—244 (1938).
— — On 14 new yeast types, produced by hybridisation. Ibid. **22**, 337—353 (1939a).
— — *Saccharomycodes ludwigii* HANSEN, a balanced heterozygote. Ibid. **22**, 357—371 (1939b).
—, and C. ROBERTS: A gene for diploidisation in yeasts. Ibid. **24**, 341—346 (1949)
WHITEHOUSE, H. L. K.: Heterothallism and sex in the fungi. Biol. Rev. **24**, 411—447 (1949).

Discussion

Lovett: I wonder in your scheme whether you can explain how mating-type shifts occur during ordinary mitotic division. Can you have recombination here?

Ahmad: In this situation I think we would have to invoke some sort of mutation.

Parag: The main difficulty in understanding heterothallism in yeast is the high frequency of the mutation from *A* to *a* and vice versa, or from heterothallism to homothallism. How can your model account for this?

Ahmad: Although this may not be true for all cases, it is still possible that in some cases the changes are due to mutations.

Snyder: We've searched in *Hypomyces* for such a mutation and we haven't found it. Have others?

Snider: Dr. Parag's comment is well taken, as the frequency of presumed spontaneous "mutants" is sufficient to favor some kind of recombination for their origin rather than mutation. In the somewhat similar history of research with Basidiomycetes, the origin of mating-type "mutants", as interpreted by Kniep, was eventually shown by Papazian to be an instance of recombination.

If the integration of the whole sexual process in yeast is combined into one operon, it is difficult to explain the indeterminate diploid vegetative stage. All the structural genes of an operon are, at a given time, typically either on or off; consequently, at least 2 operons would be required.

Ahmad: The model proposed covers only those stages of sexual reproduction which enable the organism to regain its diploid state. Meiosis is likely to be controlled by another set of genes.

The function of the mating-type locus in filamentous Ascomycetes

by

G. N. Bistis

Heterothallism in the Euascomycetes has been of interest to mycologists ever since its discovery by Dodge (1920) in *Ascobolus magnificus*. In this classic paper, he showed that heterothallism in this species is a genetically determined condition in which the two mating groups (*A, a*) are deter-

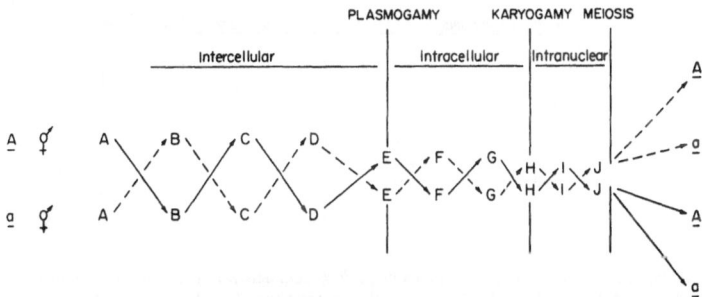

Figure 1. The function of the mating-type locus in a hypothetical heterothallic Euascomycete. The two mating classes, *A, a*, are hermaphroditic but self-incompatible. The mating-type alleles, *A, a*, determine the specificity of pairs of substances that regulate each of the steps, *A, B, C*...etc., of the sexual process. As a consequence, only the inter-class reactions occur at each of these steps. This assures the heterozygosity of the fusion nucleus in the ascus and the subsequent segregation of the mating-type alleles at meiosis. (After Raper, 1960)

mined by the two alleles (*A, a*) at a single locus. About a decade later, Gwynne-Vaughan and Williamson (1932) proved that the two mating groups are not the opposite sexes; that is, all strains are hermaphroditic.

This condition of two mating classes superposed upon hermaphroditism is now regarded as the general pattern in the heterothallic Euascomycetes (Catcheside, 1951; Raper, 1954).

Because of these two aspects, four interactions are possible in the fertile $A \times a$ mating. These are the two intra-class reactions, $a \, \male \times a \, \female$ and $A \, \male \times A \, \female$, and the two inter-class reactions, $a \, \male \times A \, \female$ and $A \, \male \times a \, \female$. The intra-class reactions, however, are never consummated, for no ascus containing spores of just one mating type has ever been reported. Apparently, within the physiological environment determined by this locus,

BOMBARDIA LUNATA

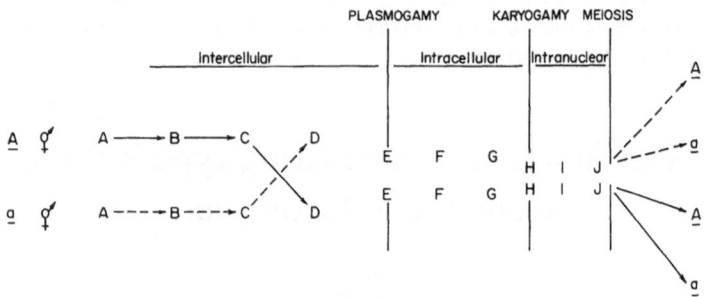

Figure 2. The function of the mating-type locus in *Bombardia lunata*. The mating-type alleles control the specificity of the substances produced by spermatia that attract trichogynes of opposite mating type (step C → D) (Zickler, 1952). All prior steps occur in isolated, single-spore cultures: nothing is known of the factors controlling succeeding steps

PODOSPORA ANSERINA

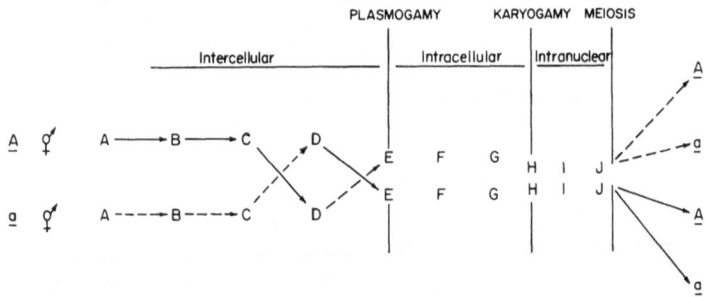

Figure 3. The function of the mating-type locus in *Podospora anserina*. The mating-type alleles control the two steps, attraction of the trichogyne to the spermatium (C → D), and plasmogamy (D → E) (Esser, 1959). All prior steps occur in isolated, single-spore (uninucleate) cultures: nothing is known of the factors controlling succeeding steps

only the inter-class reactions are completed. The problem confronting a mycologist interested in heterothallism is simply: Why is this so? This question, however, has not been fully answered as yet for any species.

Before any discussion of the known mechanisms, a theoretical model will be considered briefly (fig. 1) (Raper, 1960). In this model, every step in the sexual process is controlled by a pair of specific substances that promote only the inter-class reactions. The control is a very rigid one and includes intercellular (preplasmogamy), intracellular (postplasmogamy), and intranuclear (postkaryogamy) reactions.

In view of the diversity of sexual processes in these fungi, variations on this model would be expected. Indeed, among the three species studied, two pyrenomycetes and one discomycete, such variation has been found.

The first such study is that of ZICKLER (1952) in the eight-spored pyrenomycete, *Bombardia lunata* (fig. 2). In this species, the sexual structures arise in single-spore culture. The usual fertilizing element is a spermatium, although young spermogonia may also so function; the receptive organ is an ascogonium with a widely-ranging trichogyne. The discrimination due to mating-type occurs at a relatively late stage, namely, the attraction of the trichogyne. Spermatia produce a diffusible substance that specifically attracts only trichogynes of opposite mating-type. Except for a rare, chance encounter, this mechanism allows only the inter-class reactions to proceed to plasmogamy and eventual production of spores. The question remains, however: Is there a later block in *B. lunata* that prevents the rare intra-class encounters from proceeding to spore production?

ESSER (1959) did find such a block in the closely related four-spored species, *Podospora anserina* (fig. 3). In cultures derived from uninucleate ascospores, he found two blocks to self-fertility, one at trichogyne-attraction, the other at plasmogamy. This latter discrimination is the ideal method of control for those fungi in which a detached, transportable spermatium is sought out by a wide-ranging trichogyne from a preformed ascogonium.

The third study (BISTIS, 1956a, 1956b, 1957; BISTIS and RAPER, 1963) has been with the discomycete, *Ascobolus stercorarius*. In this species, asexual spores, oidia, are produced on all mycelia. Although after stimulation these can function as antheridia, often both sex organs, the ascogonium and the antheridium, are attached to a mycelium. As a consequence of this attachment and the limited range of the trichogyne of the ascogonium, the two sexual organs differentiate only in $A \times a$ matings at about the same time and in the same general area. The mating-type alleles therefore probably control the induction and interaction of these organs.

In order to study these early interactions, the agar-block technique was devised. The "cross" is made by placing a small block of agar bearing a detached element, such as a non-germinated oidium or a severed hypha, on the surface of a compatible four-day-old mycelium. Within 6—10 hours, the apices of the severed hypha differentiate into antheridia (fig. 5).

The principal steps and time-intervals in the sexual process of *A. stercorarius* are presented in fig. 5—8: phase I — induction of antheridia on the severed hypha — several hours; phase II — proliferation of ascogonial hyphae, induction and maturation of ascogonia on the mycelial parent — several hours; phase III — directed growth of the trichogyne and plasmogamy — ca. $^1/_2$ hour; phase IV — growth and maturation of the apothecium — ca. 7 days.

Mating-type plays a critical role at phase I. Does it play a role in any subsequent steps? Is the absence of antheridia the only impediment to completion of the intra-class reactions?

To answer these questions, substitution experiments were performed. Antheridia, induced by a compatible mycelium, were transferred to mycelia of like mating-type at various stages of sexual development. In this way the combinations $a \, \male \times a \, \female$ and $A \, \male \times A \, \female$ could be tested. The results were as follows:

1) When antheridia were transferred to vegetative mycelium there was no detectable response.

2) When antheridia were transferred to a mycelium at phase III, the trichogyne grew to and fused with the antheridium. The fruiting bodies that began to develop around these illegitimate fusions, however, always stopped growing within 24 hours, and they never produced asci.

These results can be interpreted from two points of view: 1) The mating-type alleles control only certain stages (e. g., phases I, II and IV) of the sexual process; other stages (e.g. phase III) are controlled by other loci. From this view, the involvement of mating-type in the sexual process is discontinuous. 2) These alleles function merely as activators to initiate and maintain sexual activity. Although the sexual progression itself is intact and identical for both mating-types, sexual activity occurs only in $A \times a$ combinations. When the interacting A and a elements are separated, such activity ceases within a short time. After such separation, a short-term process, such as trichogyne-attraction and plasmogamy, could show a positive response in intra-class confrontations, whereas long-term processes such as ascogonial induction and maturation of the apothecium would not.

The situation is, however, yet more complicated than this. Throughout this study, many minor differences between the two reciprocal, compatible reactions were encountered. In general, the $A \, \male \times a \, \female$ interaction appears to be the stronger. It occurs under a wider range of conditions, is more rapid, and extends over a greater area. To determine whether these differences are a reflection of a difference at the first stage in the reaction, an experiment with filtrates was performed. On the assumption that antheridial induction is the first step in sexual reproduction, severed hyphae were placed in filtrates obtained from isolated mycelia of each mating-type. Antheridia developed only on A fragments exposed to secretions of a mycelia, i.e., a mycelia in isolation produce a substance(s) capable of inducing antheridia on A severed hyphae. No antheridia appeared in the reciprocal inter-class combination, i.e., a fragment $\times A$ mycelium, nor in either of the intra-class combinations (BISTIS and OLIVE, unpublished). Thus, A mycelia in isolation either produce a very unstable antheridial inducer, or they produce no inducer at all. This latter alternative imposes the intriguing requirement that the a strain must always initiate the interaction, whether it subsequently differentiates as the antheridial or as the ascogonial parent.

Although these experiments have begun to unravel the role of the mating-type factors in $A.$ *stercorarius*, the picture is far from complete. It may not be premature, however, to try to relate these observations to the natural life cycle. In $A.$ *stercorarius* the details of the life cycle in nature (fig. 4) are well known (JANCZEWSKI, 1871; DOWDING, 1931; SEAVER, 1942). Within this life history, the relations between mating-type and sexual reproduction may be as follows: 1) In mixed cultures, because of the greater effectiveness of the *a* strain as an antheridial inducer, the interaction $A \circ\!\!\!\!\!\!\times \times a \,\female$ probably predominates[1]. 2) In single mating-type cultures, however, the basic hermaphroditism of the species can be expressed. When mites or insects carry over some compatible oidia, the difference in mass between the

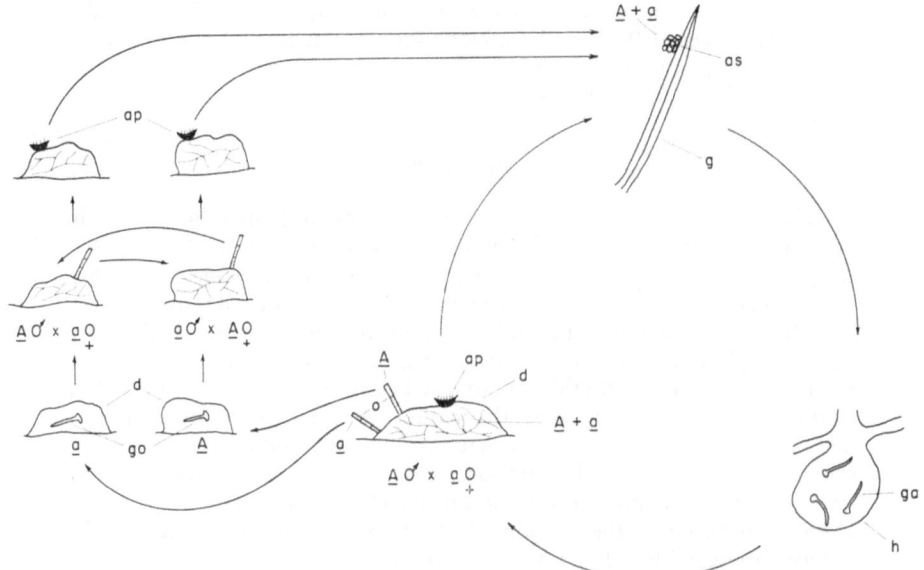

Figure 4. The life history of *Ascobolus stercorarius*. The essential features are: 1) The ascospores (as) are ejected from the ripe apothecia (ap) as octads and adhere to grass (g) or other vegetation. 2) They are injested by an herbivorous animal and germinate (ga) in the gut of the animal (h). 3) The many germinated spores in the fresh dung (d) rapidly develop into a mixed mycelium which produces apothecia (probably mainly from the interaction $A \circ\!\!\!\!\!\!\times \times a\,\female$). 4) This mycelium also produces oidia (o) that can be transported by mites or insects to uninhabited or previously colonized dung. 5) In the first instance, the oidia germinate (go) and colonize the new substrate: in the second, if the resident mycelium is of opposite mating type, they can act as fertilizing elements. In this way all mycelia can be utilized to support the development of apothecia (ap)

oidia and the mycelium insures their differentiation into oidia $\circ\!\!\!\!\!\!\times$ mycelium $\female$. No mycelium needs to be sacrificed solely to the production of asexual spores.

From this brief examination of these three examples, namely *Bombardia lunata*, *Podospora anserina*, and *Ascobolus stercorarius*, it appears that the control that the mating-type locus exerts on the sexual process of a hetero-

---

[1] The constitution of apothecia produced by mixed mycelia $(A+a)$ in the laboratory has been analyzed by means of a transplantation test. The results indicate that in most mixed mycelia of these particular strains most of the apothecia are derived from the interaction, $A \circ\!\!\!\!\!\!\times \times a\,\female$ (BISTIS and OLIVE, unpublished). The basis for this tests: apothecia, transferred to another mycelium, will continue to grow only if the recipient mycelium is of the same clone as the ascogonial parent (BISTIS, unpublished).

thallic Euascomycete is well suited to the mechanics of sexual reproduction and to the general life history of that species. This close fit suggests a long evolutionary history in which the three aspects of mating-type function, sexual mechanism, and life history have gradually evolved in each species into a well co-ordinated unit. Therefore, if we are ever to understand the origins and relationships of the various kinds of heterothallism, we need coordinated studies of these three aspects of many species. These studies should also eventually include interactions in inter-race and inter-specific matings. The mating-type alleles do not function in a vacuum but in relation to an organism and its life history.

Summary

The relation between the function of the mating-type alleles and the life cycle of three heterothallic ascomycetes, *Bombardia lunata*, *Podospora anserina*, and *Ascobolus stercorarius*, is a highly coordinated one. In the first two species, the sexual mechanism is of the transportable spermatium-preformed ascogonium type and the discrimination due to mating-type occurs at plasmogamy and/or related stages. In *A. stercorarius*, the sexual mechanism is of the antheridium (attached or transportable) — induced ascogonium type. In this instance, the mating-type alleles function at antheridial induction, the first step in sexual reproduction, and also at subsequent stages. It is not clear whether the control is a continuous or a discontinuous one. A further complication is the greater effectiveness of the antheridial inducer produced by the *a* strains. In fact, the evidence for an antheridial inducer produced by *A* strains is all indirect.

References

BISTIS, G. N.: Studies on the genetics of *Ascobolus stercorarius*. (BULL.) SCHRÖT. Bull. Torrey Botan. Club **83**, 35—61 (1956a).
— Sexuality in *Ascobolus stercorarius* I. Morphology of the ascogonium; plasmogamy; evidence for a sexual hormonal mechanism. Am. J. Botany **43**, 389—394 (1956b).
— Sexuality in *Ascobolus stercorarius* II. Preliminary experiments on various aspects of the sexual process. Am. J. Botany **44**, 436—443 (1957).
—, and J. R. RAPER: Heterothallism and sexuality in *Ascobolus stercorarius*. Am. J. Botany **50**, 880—891 (1963).
CATCHESIDE, D. G.: The Genetics of Micro-organisms. New York: Pitman 1957.
DODGE, B. O.: The life history of *Ascobolus magnificus*. Mycologia **12**, 115—134 (1920).

→

Figure 5—8. Four phases in the sexual process of *Ascobolus stercorarius*. All photographs are of the reaction *A* ♂ × *a* ♀. Figure 4. — Phase I — Induction of antheridia (an) on the severed hypha by the secretions of the four-day-old mycelium. Figure 5. — Phase II — The reaction of the mycelial parent to the antheridium (an): Proliferation of ascogonial hyphae (a.h.), the differentiation of ascogonial primordia (a.p.), and their development into ascogonia (as.). Figure 6. — Phase III — Directed growth of the trichogyne (tr) to the antheridium (an) and plasmogamy. Figure 7. — Phase IV — Growth and maturation of the resultant apothecium. Figure 4, 5 and 7 = 100×. Figure 6 = 200×

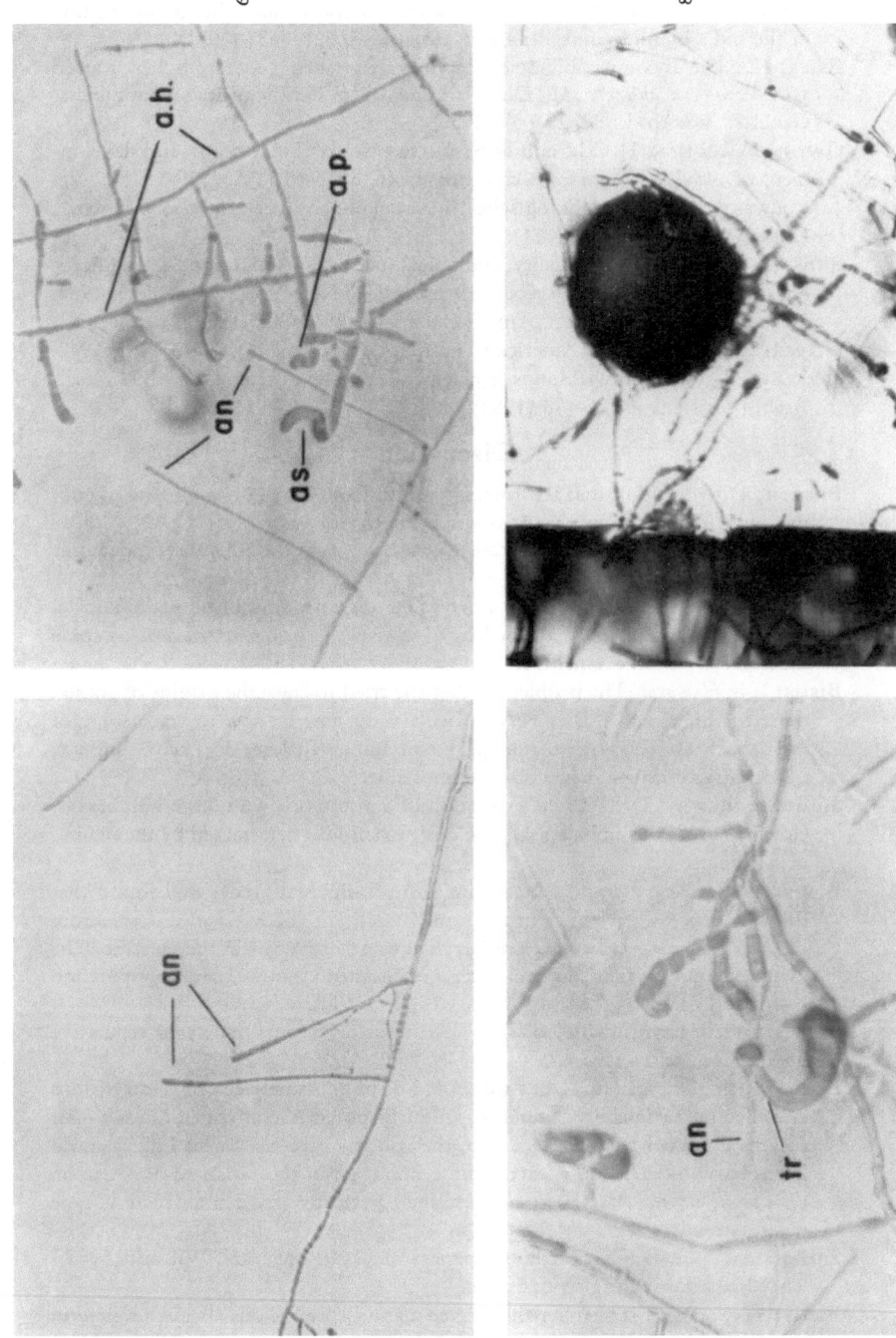

Dowding, E. S.: The sexuality of *Ascobolus stercorarius* and the transportation of the oidia by mites and flies. Ann. Botany **45**, 621—638 (1931).

Esser, K.: Die Incompatabilitätsbeziehungen zwischen geographischen Rassen von *Podospora anserina*. II. Die Wirkungsweise der Semi-Incompatibilitäts-Gene. Z. Vererb.-L. **90**, 29—52 (1959).

Gwynne-Vaughan, H. C. I., and H. S. Williamson: The cytology and development of *Ascobolus magnificus*. Ann. Botany **46**, 653—670 (1932).

Janczewski, E. G.: Morphologische Untersuchungen über *Ascobolus furfuraceus*. Z. Botan. **29**, 256—262 (1871).

Raper, J. R.: Life cycles, sexuality, and sexual mechanisms in fungi. Sex in Microorganisms: 42—81. Am. Assoc. Advance. Sci., New York, New York 1954.

— The control of sex in fungi. Am. J. Botany **47**, 794—808 (1960).

Seaver, F. J.: The North American cup-fungi (Operculates). New York 1942.

Zickler, H.: Zur Entwicklungsgeschichte des Askomyzeten *Bombardia lunata*. Arch. Protistenk. **98**, 1—70 (1952).

Discussion

Burnett: Do you think that the transformation from *a* to *A* is a general phenomenon in the Ascomycetes, or is it specific to *Ascobolus*?

Bistis: I don't know. No other cases have been as sufficiently investigated as *Ascobolus*.

Burnett: Is it not possible to have a heterokaryon with both incompatible nuclei as seems to have been described by Gwynne-Vaughan in *Humaria* or some other genus? I can't remember.

Bistis: It is *Humaria*. The problem is that she tried to trace the origins of anastamosed hyphae, and this is very difficult. You may be able to trace some of them back and determine that plus and minus fusions do occur, but any quantitative estimate is just about impossible.

Snider: Following Dr. Burnett's question, did you work with large numbers of clones to see if this imbalance in the basic reaction is determined by the mating genes themselves?

Bistis: Yes, we have studied 20 strains from 5 different stocks and found that this imbalance always goes in the same direction. The imbalance is therefore due to the mating type alleles rather than to other factors in the genome. The evidence, then, is that this is a general phenomenon in the natural population -----or so I think.

Snider: What happens in *Ascobolus* when fusion occurs between vegetative hyphae of *A* and *a* strains?

Bistis: Again, I don't know, but we should distinguish between physical fusions and effective fusions. In general -----in the 8-spored Ascomycetes, at least-----if fusions do occur between *A* and *a*, they do not give rise or lead to balanced heterokaryons. The only exception I know of is the work of El-Ani on *Ceratostomella radicola*. Here he described perfectly good inter-mating type heterokaryons.

Snider: Are you familiar with the papers of Garnjobst and Wilson (1956)? They had inter-mating type fusions.

Bistis: Yes, but as I remember these inter-mating type fusions always resulted in the death of the cells that fused. No stable heterokaryon resulted.

Snider: Agreed. The distinction seems important, however, as it places the site of specific action in *Neurospora* within the cytoplasm rather than at or in the cell wall.

The genetics of tetrapolar incompatibility

by

P. R. Day

In the tetrapolar Basidiomycetes, incompatibility is observed when two haploid, homokaryotic mycelia meet each other. The two mycelia will either react to form a dikaryotic mycelium, composed largely of binucleate cells, which is capable of forming a fruit body under suitable conditions, when they are said to be fully compatible, or they will not. The products of the latter, or incompatible, reaction are of several kinds and will be discussed in a moment. In the compatible reaction, the dikaryon is established as a result of the migration of nuclei through preformed hyphae (BULLER, 1931; SNIDER, 1963). In the majority of forms, it appears as new growth from such hyphae. The dikaryon is characterized by altered branching patterns, the possession of a clamp connection at each cross wall, its genetical properties as a special kind of heterokaryon, and its ability to form fruit bodies in which meiosis takes place.

The tetrapolar basidiomycetes derive their name from the fact that the progeny of a single fruit body are of four different mating-types. Each of the four kinds of haploid homokaryotic mycelia is only compatible with one other among the remaining three.

When the progenies of two independently produced fruit bodies are compared, it is usual to find that the four groups of each are fully intercompatible; that there are 8 mating-types. The facts can be readily accounted for by a genetic control in which mating-type is determined by two independent loci, *A* and *B*. At each of the two loci there is a multiple allelomorphic series.

We denote the haploid stocks as *A1 B1*, *A2 B2* ... etc. where the alleles are shown as numbers. The four mating-types from the dikaryon (*A1 B1* + *A2 B2*) are *A1 B1*, *A1 B2*, *A2 B1* and *A2 B2*, which occur in equal frequencies because of random assortment at meiosis. A stock which is *A1 B1* is compatible with *A2 B2* but not with any other stock which carries a common *A* or *B* allele (*A1 B2*, *A2 B1*, and *A1 B1*). As an outbreeding mechanism the tetrapolar incompatibility system is an improvement on the bipolar system with multiple alleles at only one locus. Only 25% of random sib matings are fertile, whereas 50% are fertile in bipolar forms. In both systems, as the number of alleles in the population becomes larger, the probability that non-sister matings are compatible rises rapidly.

The extent of the multiple allelomorphic series has now been adequately gauged for one tetrapolar form, *Schizophyllum commune*, by RAPER and his group (RAPER, KRONGELB and BAXTER, 1958). On the basis of a world-wide sample of 114 homokaryons, estimates of 339 (with 5% probability limits of 216 and 562) A alleles and 64 (with 5% limits of 53 and 79) B alleles in the natural population of $S.$ *commune* were made.

A. Distinction in function between the A and B loci

The incompatible mating reactions are of great interest since by examining them we would expect to find some clues to the mechanism of incompatibility.

Three kinds of incompatible reaction are possible; these are due to common A, common B or both. In each of these the product of interaction can be shown to be a heterokaryon. The common A heterokaryon is the most readily produced and is formed by the migration of nuclei comparable to, but perhaps not quite so extensive as, that which takes place in a fully compatible mating to form the dikaryon. Unlike the dikaryon, however, the common A heterokaryon does not possess clamp connections, and the ratio of the two types of nuclei in the mycelium may vary widely. The common B heterokaryon is produced only at the junction of the two mated homokaryons. Very little nuclear migration takes place. The mycelium possesses false clamp connections which resemble the true clamps of a dikaryon, except that they fail to fuse with the cell behind. Thus, only the tip cells are heterokaryotic and the heterokaryon readily breaks down unless forced on minimal medium with the aid of auxotrophic mutants. The common AB heterokaryon is even less stable and can only be grown as a forced heterokaryon. In general, under normal conditions, none of the heterokaryons from incompatible matings, nor the homokaryons, will fruit.

Two striking features emerge from these observations:

1. Nuclear migration only takes place when mycelia with different B's are paired.
2. Clamps or false clamps are only formed when mycelia with different A's are paired.

Thus the designation of mating-type factors can no longer be arbitrary, since A controls clamp formation and B controls nuclear migration. This distinction in function between A and B has now been observed in 11 different species (FULTON, 1950; SWIEZYNSKI and DAY, 1960; TAKEMARU, 1961).

Additional information on the role of the B factor is presented on p. 70.

B. Structure of the mating-type loci

More than 40 years ago, KNIEP (1923) reported that the progenies of single fruit bodies of $S.$ *commune* would sometimes include more than four mating-types. Intensive work during the last 13 years has shown that the appearance of new mating-type factors is not due to mutation, as was first proposed, but is the result of recombination between component loci of the A and B loci (PAPAZIAN, 1951; TAKEMARU, 1957, RAPER, BAXTER and MIDDLETON, 1958; RAPER, BAXTER and ELLINGBOE, 1960; TERAKAWA, 1960; DAY, 1960; DAY, 1963b).

To take the A mating-type "locus" as an example, it seems that this is now better defined as two loci, or perhaps a super-gene, composed of two loci, $A\alpha$ and $A\beta$. Results from the two Hymenomycetes *Schizophyllum commune* and *Coprinus lagopus* agree in most respects. In place of *A*1, we may now write the alternative notation *A*1 *a*1—*β*1, and when this factor meets *A*2 *a*2—*β*2 in a cross and recombination occurs, the recombinants may be written *Ax a*1—*β*2 and *Ay a*2—*β*1, where *x* and *y* may, or may not, correspond to number specifications of wild factors in the population. *Ax* and *Ay* behave as fully functional alleles which are self incompatible but are cross-compatible with the original parental alleles and with each other. In the latter type of cross, the original parental specificities, *A1* and *A2*, are regenerated by recombination. Since both recombinant *A* factors are compatible with both parental factors, we can conclude that two *A* factors only need differ by alleles at one locus, either *Aa* or *Aβ*, in order to be compatible. This fact is of course of some significance for any theory which attempts to explain the mode of action of the mating type factors.

Schizophyllum and *Coprinus* differ rather markedly in that recombination between *Aa* and *Aβ* in *Schizophyllum* is approximately a hundred times more frequent than it is in *Coprinus*. In *Schizophyllum*, there is also a marker (*paba*) located between *Aa* and *Aβ*. The discovery of the *Aa* and *Aβ* loci necessitated a re-examination of the data on factor frequencies in wild populations. This has been done in *Schizophyllum* (RAPER, BAXTER and ELLINGBOE, 1960) and, because the analysis is more tedious, to a lesser extent, in *Coprinus* (DAY, 1963b). In both fungi, estimates of the number of *Aa* and *Aβ* alleles have been made, and both are similar in having more *Aβ* specificities than *Aa* ones.

All attempts to find intra-locus recombination which might be expected to generate new functional or other novel specificities have so far failed. Even with present techniques, it is very tedious to push the level of analysis beyond recovery of products at the rate of 10^{-5}, and it would seem that this level of resolution is probably not fine enough.

The *B* locus of *Schizophyllum* appears to be more complex than the *A* locus in that, while recombination occurs, the specificities generated in some crosses cannot easily be made to fit a 2 locus nor yet even a 3 locus model (RAPER, pers. comm.). In *Coprinus*, no evidence for recombination at the *B* locus has been obtained. In both fungi, the resolution of the structure of the *B* locus is severely handicapped by the lack of suitable *B*- linked markers despite intensive efforts to find them.

C. Mutation of the mating-type loci

The large numbers of alleles at the *A* and *B* loci of some of the intensively investigated forms raises the question of how they originated. While the earlier workers often attributed deviant forms to mutation rather than to recombination, or contamination, it is only recently that this aspect of fungal incompatibility began to be thoroughly explored.

The nature of the mating-type loci suggests that the most efficient means of screening mutants would be by exploiting the differences in properties between dikaryons arising by mutation in common *A* or common *B* heterokaryons and the heterokaryons themselves. In *Coprinus* and *Schizophyllum*, this was done by making use of the fact that heterokaryons rarely ronever fruit. In large scale inoculations of treated and untreated heterokaryons of both fungi to suitable fruiting media, the rare fruit bodies that did arise proved to originate directly as a result of mutations in the hetero-karyons.

The mutants of most interest proved to be those of the *A* factor in *Coprinus* (SWIEZYNSKI and DAY, 1960; DAY, 1963a) and of the *B* factor in *Schizophyllum* (PARAG, 1962). The *A* factor mutants in *Coprinus* were of interest in two respects. First they were mutants which lacked mating-type specificity: not only were they compatible with all available tester stocks but they were self-compatible. Secondly, in homokaryons, they produced a phenotype not unlike that of a common *B* heterokaryon. The hyphae pro-duced false clamps at many of the cross walls. This supports the conclusion that the *A* factor controls the formation of clamp connections. Evidently the interaction normally required between two *A* factors, different by at least one allele, before clamps were formed was in some way by-passed in the mutants. Four out of 10 such mutants were tested for functional *A*α and *A*β alleles. All had functional *A*α alleles which could be recovered in recombinants. None of them had functional *A*β alleles which could be recovered in this way.

The *B* factor mutants in *Schizophyllum* proved to be similar. These mutants in homokaryons, as one might predict, mimicked the character-istic morphology of the common *A* heterokaryon of *Schizophyllum*. Like the *A* mutants of *Coprinus*, they proved to have functional *B*α alleles, but no functional *B*β alleles could be recovered. It should be pointed out that at the present time, the designation α and β for the *A* and *B* sub units of both *Coprinus* and *Schizophyllum* is quite arbitrary and of itself implies no similarity between the two fungi. Indeed, so far there is no evidence of a functional difference between *A*α and *A*β in either fungus.

RAPER (pers. comm.) has recently been able to isolate *A* mutants in *Schizophyllum* which closely resemble those of *Coprinus*.

D. Suppressors and modifiers

In normal wild type stocks and in the mutants described so far, the control of mating-type appears to be firmly vested in the mating-type loci. However, as one might expect, other loci have now been identified which modify the control exercised by the mating-type loci. A simple example of this is the mutant *su-A*, which segregates independently of the *A* and *B* loci in *Coprinus* (DAY, 1963a). This mutant was isolated from a fruit body formed when a common *A* heterokaryon was inoculated to fruiting medium. *su-A* is a recessive *A* factor suppressor, which, in a homokaryon, gives a phenotype very much like that of an *A* mutant. The mycelium has many false clamps but, unlike the majority of *A* mutants, will readily form fruit bodies. These have relatively high frequencies of 1, 2 and 3 spored basidia. The *A* factor suppressor is not allele specific.

In *Schizophyllum*, some 9 modifiers which affect heterokaryosis have also been found, but they are entirely different from the *A* suppressor in *Co-prinus* (RAPER and RAPER, 1964). None of these mutations is phenotypically expressed in the homokaryon. However, in common *A* heterokaryons, in homokaryons carrying a mutant *B* factor, or in disomics heteroallelic for *B*, the mutations are expressed. They bring about the formation of false clamps. Even more remarkable is the fact that 8 of these mutations also modify the phenotype of dikaryons and probably that of common *B* heterokaryons. All of these modified mycelia have common phenotypic characteristics not found in any of the normal heterokaryons.

When homokaryons were isolated from such heterokaryons, the phenotype of the parent culture usually persisted through several successive subcultures but never persisted through fruiting.

The phenotype varies with the dosage of the modifier, a more extreme phenotype being found in heterokaryons carrying two modifiers, one in each nucleus, than in heterokaryons with only a single dose.

As we have seen, the formal genetics of the tetrapolar Basidiomycetes has revealed a number of interesting problems which cry out for further investigation. The most striking of these is the nature of the gene action involved which I shall leave to the speakers who will follow.

E. Summary

An account is given of the different functions of the *A* and *B* loci, their genetic fine structure, mutational changes and the effects of suppressors and modifiers. The examples used are taken largely from work on *Schizophyllum commune* and *Coprinus lagopus*.

References

BULLER, A. H. R.: Researches on Fungi. IV. Longmans, Green & Co. London. 1931.

DAY, P. R.: The structure of the A mating type locus in *Coprinus lagopus*. Genetics **45**, 641—650. (1960).

— Mutations of the A mating type factor in *Coprinus lagopus*. Genet. Res. **4**, 55—64 (1963a).

— The structure of the A mating type factor in *Coprinus lagopus:* Wild alleles. Genet. Res. **4**, 323—325 (1963b).

FULTON, I. W.: Unilateral nuclear migration and the interactions of haploid mycelia in the fungus *Cyathus stercoreus*. Proc. Natl. Acad. Sci., Wash. **36**, 306—312 (1950).

KNIEP, H.: Über erbliche Änderungen von Geschlechtsfaktoren bei Pilzen. Z. ind. Abst.- u. Vererb.-Lehre **31**, 170—183 (1923).

PAPAZIAN, H. P.: The incompatibility factors and a related gene in *Schizophyllum commune*. Genetics **36**, 441—459 (1951).

PARAG, Y.: Mutations in the B incompatibility factor of *Schizophyllum commune*. Proc. Natl. Acad. Sci., Wash. **48**, 743—750 (1962).

RAPER, C. A. and J. R. RAPER: Mutation affecting heterokaryosis in *Schizophyllum commune*. Am. J. Botany **51**, 503—512 (1964).

RAPER, J. R., M. G. BAXTER, and A. H. ELLINGBOE: The genetic structure of the incompatibility factors of *Schizophyllum commune:* the A factor. Proc. Natl. Acad. Sci., Wash. **46**, 833—842 (1960).

— —, and R. B. MIDDLETON.: The genetic structure of the incompatibility factors in *Schizophyllum commune*. Proc. Natl. Acad. Sci., Wash. **44**, 889—900 (1958).

—, G. S. KRONGELB, and M. G. BAXTER: The number and distribution of incompatibility factors in *Schizophyllum*. Am. Naturalist **92**, 221—232 (1958).

SNIDER, P. J.: Genetic evidence for nuclear migration in Basidiomycetes. Genetics **48**, 47—55 (1963).

SWIEZYNSKI, K.M. and P.R. DAY: Heterokaryon formation in *Coprinus lagopus*. Genet. Res., Camb. **1**, 114—128 (1960).

TAKEMARU, T.: Genetics of *Collybia velutipes*. V. Mating patterns between F1 mycelia of legitimate and illegitimate origins in the strain NL-55. Botan. Mag. (Tokyo) **70**, 244—249 (1957).

— Genetical studies on fungi. X. The mating system in Hymenomycetes and its genetical mechanism. Biol. J. Okayama Univ. **7**, 133—211 (1961).

TERAKAWA, H.: The incompatibility factors in *Pleurotus ostreatus*. Sci. Papers Coll. Gen. Educ. Univ. Tokyo **10**, 67—71 (1960).

Discussion

Bistis: Is there any general rule with respect to the constitution of A factors and subunit specificity? When does an A factor change? When do you get a new A factor? Must two different A factors be different at both subunits, or will difference at one suffice?

Day: The results of crossing two A factors which differ in both subunits indicate that the recombinants formed ---- each of which differs from the parents in only one subunit -----are compatible with both parents. From this we can conclude that factors need to differ at only one subunit to be compatible with one another. We do not know how new, functional, subunit specificities arose during the course of evolution. The only laboratory mutants recovered so far appear to be defective.

Singer: How would the role of the A factor be interpreted in such genera where clamp formation does not take place. They could, with A controlling the clamp formation, not be considered as tetrapolar. Are they all homothallic?

Day: I thought that most are not tetrapolar, but that in any that are bipolar, the mating type factor would be analogus to the B factor of tetrapolar forms and would only control nuclear migration.

Burnett: There are at least four tetrapolar fungi which don't produce clamps.

Day: They ought to be looked into.

Somatic recombination in Basidiomycetes

by

ALBERT H. ELLINGBOE

There is now unequivocal proof of the existence of mechanisms for genetic variability in vegetative cells. Genetic recombination has been clearly demonstrated in vegetative cells of several species of filamentous fungi. This paper will briefly review what is known of the mechanisms for genetic exchange or recombination in vegetative cells of the Basidiomycetes.

A. Uredinales

First let us consider genetic recombination in the Uredinales, the rusts, the heterothallic members of which have a 1-locus, 2-allele ("+" and "—") incompatibility system. Somatic recombination involving paired dikaryons has been reported in each of several species (fig. 1). Each of the dikaryotic strains, though heteroallelic for several genes, was stable and yielded no recombinants upon subculturing for several asexual (uredial) generations. When two such dikaryons were grown together as a dual infection, however, anastomoses occurred and recombinats were obtained (fig. 1). The

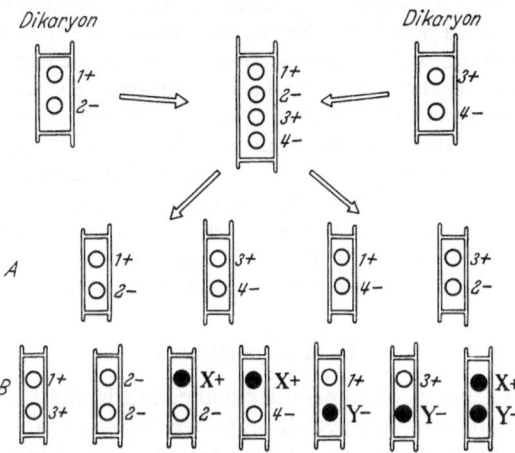

Figure 1. The dikaryotic × dikaryotic matings in *Puccinia graminis* var. *tritici* and an interpretation of the results obtained by NELSON et al. (1955) (*A*) and of WATSON (1957) (*A+B*). The parental nuclei are numbered 1, 2, 3 and 4, recombinant nuclei are labeled *X* and *Y*, mating types are designated + (plus) and — (minus)

method by which recombinational events were detected is important in interpreting the results, because the methods appear to have determined the type of results obtained and, therefore, the interpretation of the data.

In one case (NELSON et al. 1955), represented by the second line (A) in Fig. 1, the selection was for virulence on a host on which neither of the parental dikaryons was virulent. The purified virulent strain, when transferred to a non-selective host, sectored into 4 types — two parental types and 2 recombinant types — explainable on the basis of reassociation of whole nuclei. The new virulent strain, therefore, must have possessed all 4 parental types of nuclei. The 4 non-virulent strains derived from the virulent strain could be the result of reassociation of the 4 nuclei into compatible pairs.

Other studies (WATSON, 1957; WATSON and LUIG 1958; ELLINGBOE, 1961) used the recombination of characters (such as spore color and pathogenicity), present in the 2 parental strains, as a means of detecting recombinational events. These studies have shown that exchange of whole nuclei cannot explain the large number of phenotypes observed in the progeny (see lines A and B in fig. 1). If the original pairs of nuclei were $(1^+ + 2^-)$ in one strain and $(3^+ + 4^-)$ in the other strain, a reassociation of "plus" and "minus" nuclei would give the 2 new strains $(1^+ + 4^-)$ and $(3^+ + 2^-)$. The existence of stable trikaryons or tetrakaryons could account for the occurence of 4 more phenotypes of the constitution $(1^+ + 2^- + 3^+)$, $(1^+ + 2^- + 4^-)$, $(1^+ + 3^+ + 4^-)$, $(2^- + 3^+ + 4^-)$ and $(1^+ + 2^- + 3^+ + 4^-)$. If 2 "plus" nuclei or 2 "minus" nuclei could exist together in a cell, an additional 6 phenotypic classes could be observed, namely $(1^+ + 3^+)$, $(2^- + 4^-)$, $(1^+ + 1^+)$, $(2^- + 2^-)$, $(3^+ + 3^+)$, and $(4^- + 4^-)$. The occurrence of stable trikaryons, tetrakaryons, or dikaryons with like mating type factors is seriously questioned as a result of studies with other organisms but nevertheless is a possibility. Since recovery of more than 13 new phenotypes from a mating of 2 dikaryotic strains could not be explained soley on the basis of the exchange of whole nuclei, additional mechanisms for variability were postulated. Rather than invoking the cytoplasm as the source of phenotypic variability, it has been suggested that a "parasexual cycle", in the broadest sense, was involved, i.e., recombinant nuclei arise via the recombination of genes of 2 or more nuclei. The data on the rusts, however, do not permit a clear distinction between known mechanisms of somatic recombination. In fact, the inheritance of pathogenicity (the genetic characters) via the standard sexual cycle has not been fully analyzed.

B. Ustilaginales

Our understanding of somatic recombination in the Ustilaginales, the smuts, is far more precise, especially in *Ustilago maydis*. In *U. maydis*, mating competence and pathogenicity are determined by a 2-locus (A and B)

Table 1. *Segregation from a diploid of Ustilago maydis following alpha irradiation.*

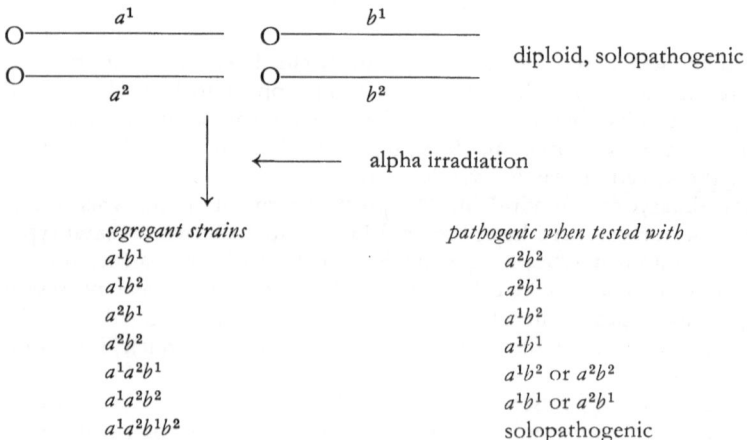

segregant strains	pathogenic when tested with
a^1b^1	a^2b^2
a^1b^2	a^2b^1
a^2b^1	a^1b^2
a^2b^2	a^1b^1
$a^1a^2b^1$	a^1b^2 or a^2b^2
$a^1a^2b^2$	a^1b^1 or a^2b^1
$a^1a^2b^1b^2$	solopathogenic

incompatibility system. Generally, the haploid phase of the smuts is saprophytic and the dikaryotic phase, parsitic. Ocassionally, however, single, uninucleate, diploid basidiospores, or sporidia, have been obtained which produced colonies on synthetic media and also were pathogenic on a host plant, i. e., a solopathogenic strain (CHRISTENSEN, 1931). Diploids have also been obtained by isolating prototrophs from diseased host plants which had been inoculated with 2 complimentary auxotrophic strains (HOLLIDAY, 1961).

Two types of segregation have been observed from diploids of *Ustilago maydis*, namely, crossing over and haploidization. Recombination of nutritional requirements in a diploid could best be explained by mitotic crossing-over (HOLLIDAY, 1961). None of the recombinants was haploid as determined by pathogenicity tests. Furthermore, segregants from the first

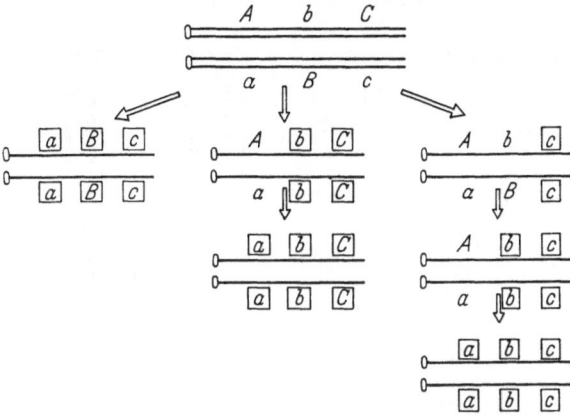

Figure 2. Segregation from a diploid of *Ustilago maydis* when selection was for the *a* (left), the *b* (middle), or the *c* (right) phenotype. The loci for which the diploid is recombinant (i. e., now homozygous) are presented in boxed letters

recombinational event, when again subjected to conditions favorable for recombination, showed further segregation. Fig. 2 presents some of the types of segregation from a diploid. When selection was for the recessive *a* phenotype, the *c* phenotype was also present, and further subculturing gave no recessive *b* phenotype which would have indicated heterozygosity for *Bb*. Selection for the *b* phenotype gave individuals which were homozygous for *C* and only some of which were heterozygous for *Aa*. Some individuals selected with the *c* phenotype showed further segregation in subsequent generations for both the *a* and *b* loci. The somatic recombination observed in *Ustilago maydis* was, therefore, consistent with mitotic crossing over as described in Drosophila by STERN (1936) and in *Aspergillus nidulans* by PONTECORVO (1956).

Treatment of a diploid, $a^1a^2b^1b^2$, with alpha irradiation caused an apparent haploidization (tab. 1) (ROWELL, 1955). Isolates selected either at random or because of abnormal morphology were tested for their pathogenicity by themselves and with tester strains. One strain was inferred to be a^1b^1 because it was pathogenic when mated with tester a^2b^2. Another was inferred to be a^1b^2 because is was pathogenic when mated with tester

a^2b^1. Another strain was inferred to be disomic for a, that is, $a^1a^2b^1$, because it was pathogenic when mated with either of 2 testers, a^1b^2 or a^2b^2.

The mechanism of somatic recombination in the smuts appears to be essentially identical to the Parasexual Cicle described by PONTECORVO (1956) in *Aspergillus nidulans*, although, a clear demonstration that haploidization reassociates whole chromosomes is wanting.

C. Hymenomycetes

Somatic recombination in the tetrapolar Hymenomycetes appears to be more diversified, possibly it may be suggested, because of the numerous methods used to detect recombinational events.

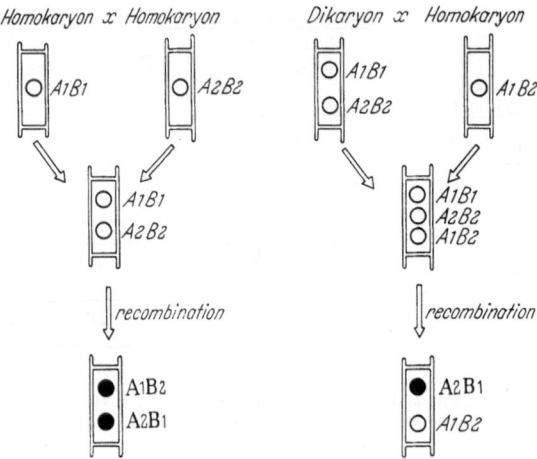

Figure 3. Two types of matings in which recombinational events have been detected in tetrapolar hymenomycetes. Mating-type constitution is presented adjacent to the nucleus. Mating-type factors of recombinant nuclei are in bold faced letters

Two types of matings have been made: (1) homokaryon × homo-karyon ⟶ dikaryon or heterokaryon (recombination) ⟶ new dikaryon or new heterokaryon, (2) dikaryon × homokaryon ⟶ new dikaryon as a result of recombination of mating-type factors and subsequent dikaryotization of homokaryon (fig. 3).

Although the data were very limited, the single reported study of the first type (PARAG, 1962) indicated recombination in the dikaryon in the absence of a homokaryotic mate to proceed via a mechanism in which transient diploid nuclei underwent a process of haploidization that was similar to meiosis in that reassociation of linked genes was frequent.

The use of dikaryotic-homokaryotic matings to detect recombinational events has produced varied results. One report (CROWE, 1960), on recombination in fully compatible matings of *Schizophyllum commune* (fig. 4), i. e. $(A1\ B1 + A2\ B2) \times A3\ B3 \longrightarrow A1\ B1 + A3\ B3,\ A2\ B2 + A3\ B3,$ $A1\ B2 + A3\ B3,$ or $A2\ B1 + A3\ B3$, etc., suggested the haploidization process to be a meiosis-like event in that crossing over was common among the progeny. The sample of recombinant sectors was small, however, and the selective role of the homokaryon in the recovery of recombinant nuclei was not determined.

Recombination of mating-type factors in noncompatible dikaryotic-homokaryotic matings has been used quite extensively to detect recombinational events (fig. 5) (QUINTANILHA, 1939; PAPAZIAN, 1954; ELLINGBOE, 1963, 1964; ELLINGBOE and RAPER, 1962). In a noncompatible dikaryotic-homokaryotic mating of the type ($A1\ B1 + A2\ B2$) × $A1\ B2$, the 3 types of nuclei were brought together in a common mycelial system, and dikaryotic sectors were occassionally produced in the homokaryon. The dikaryotic sectors might have arisen in one of 2 ways (ELLINGBOE, 1963): 1) the 2 nuclei from the dikaryon displaced the homokaryon, i. e., ($A1\ B1 + A2\ B2$) × $A1\ B2 \longrightarrow A1\ B1 × A2\ B2$, or 2) a recombinational event occured which produced a nucleus that was compatible with the homokaryon, i. e., ($A1\ B1 + A2\ B2$) × $A1\ B2 \longrightarrow A2\ B1 + A1\ B2$. The formation of the dikaryotic sector of the latter type clearly requires the occurence of a recombination event.

The dikaryotic sectors which appear in the homokaryon in both compatible and noncompatible dikaryotic-homokaryotic matings are not always simple dikaryons, but are frequently functionally "trikaryotic" (ELLINGBOE, unpublished) with the sectors containing all 3 of the original types of nuclei. In *S. commune*, the use of 3 different mutations to auxotrophy, $x1$,

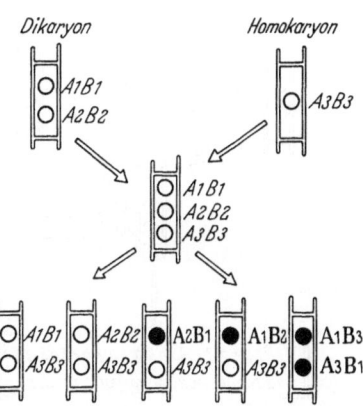

Figure 4. Types of sectors obtained in compatible dikaryotic-homokaryotic matings in *Schizophyllum commune*. Mating-type constitution is presented adjacent to the nucleus. Mating-type factors of recombinant nuclei are in bold faced letters

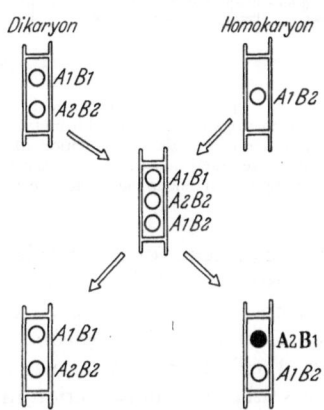

Figure 5. Types of sectors obtained in non compatible dikaryotic matings in *Schizophyllum commune*. Mating-type constitution is presented adjacent to the nucleus. Mating-type factors of recombinant nuclei are in bold faced letters

$x2$ and $x3$, in a cross ($A1\ B1\ x1\ x2 + A2\ B2\ x2\ x3$) × $A3\ B3\ x1\ x3$ on minimal medium, allowed for selection and maintenance of a functionally trikarotic mycelium (fig. 6), since all possible pairs of nuclei would have one nutritional requirement in common (ELLINGBOE, unpublished). Selection applied for wild type alleles of auxotrophic mutations in essentially this manner with *Coprinus lagopus* allowed growth on minimal media, but only a small portion of isolated hyphal tips produced colonies when transferred to minimal media (SWIEZYNSKI, 1963). Not all hyphal tips from such cultures would be expected to grow if the parent cultures were, in fact, trikaryotic, and if some of the hyphal tips did not contain all 3 types of nuclei. From each of 13 matings, a colony derived from a single hyphal tip plated on minimal medium was also fruited. Fruits from four of the 13 matings showed segregation as expected from a triploid, 2 additional showed what is also probably segregation from a triploid, 4 showed segregation from a diploid, and 3 showed what is probably segregation from a diploid trisomic for the *B*-linkage group. The same basic procedure, that is, the use of the "trikaryon", has been used to detect recombinational events in *Coprinus radiatus*, and, again, a fairly high percentage (over 30%) of the sectors showed segregation characteristic of aneuploids (PRUD'HOMME, 1963).

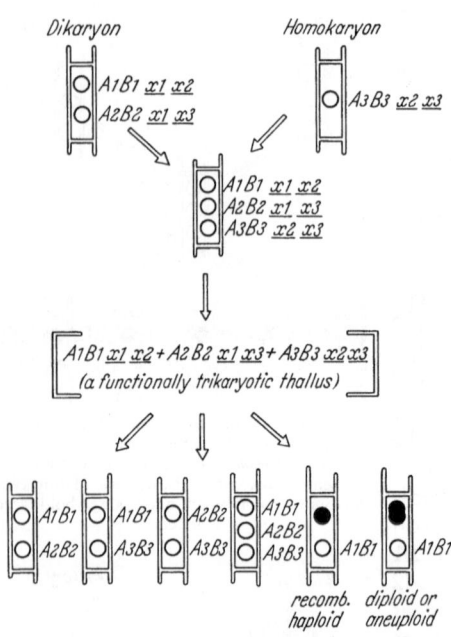

Figure 6. Types of sectors obtained in dikaryotic-homokaryotic matings when auxotrophic mutations were used for the detection of genetic recombination. Mating-type constitution is presented adjacent to the nucleus. Recombinant and/or diploid or aneuploid nuclei are presented in black

The presence of aneuploids is not a satisfactory criterion for concluding the existence of the Parasexual Cycle, as described by PONTECORVO in 1956, since aneuploids can arise via the standard sexual cycle as demonstrated in both *Coprinus* and *Schizophyllum* (SWIEZYNSKI, 1963; RAPER and OETTINGER, 1962). In fact, a regular method of producing aneuploids in higher plants utilizes nondisjunction in meiosis of structural hybrids. The mechanism for the origin of recombinant nuclei in *Coprinus* spp. is, therefore, still in question.

A different procedure has been used for the study of recombinational events in noncompatible dikaryotic-homokaryotic matings in *Schizophyllum commune*, i. e., $(A1\ B2 + A2\ B1) \times A1\ B1$ (ELLINGBOE and RAPER, 1962). Recombinational events were detected by the recombination of mating type factors, i. e., $(A1\ B2 + A2\ B1) \times A1\ B1 \longrightarrow A2\ B2 + A1\ B1$. The origin of the $A2\ B2$ nucleus was the central question in most of the studies. The mechanisms of recombination were deduced by analysis of the recombination which occured among linked and nonlinked auxotrophic mutations in the events that reassociated the mating-type factors.

When a dikaryotic sector arose in the homokaryon of the dikaryotic-homokaryotic mating, it was removed, purified by single hyphal tip iso-

Table 2. *The procedure for the determination of the incompatibility factors in each of the nuclei of a dikaryotic sector (detailed explanation in text)*

	reaction with testers			
types of dikaryotic sectors in homokaryon	A1 B1	A1 B2	A2 B1	A2 B2
(A1 B2 + A2 B1) X A1 B1 → A1 B2 + A2 B1	−	+	+	−
A2 B2 + A1 B1	+	−	−	+
Ax B2 + A1 B1	+	−	+	+
Ax B1 + A2 B2	+	+	−	+
Ax B1 + Ax B2	+	+	+	+

Table 3. *The procedure for the heterokaryotic allelic tests used to determine the auxotrophic mutations present in the recombinant A2 B2 nuclei of a dikaryotic sector (detailed explanation in text)*

(A1 B2 + A2 B1) X A1 B1 → (A2 B2 + A1 B1)

x2	x3	x1	?	x1
x4	x9	x7		x7
x5				
x6				
x8				

		media	
		minimal	complete
(A2 B2 + A1 B1) X A1 B1 → (A2 B2 + A1 B2) →		−	+
? x1 x1 x1 x1			
x7			
„ X A1 B1 → (A2 B2 + A1 B1) →		+	+
x2 x2			
„ X A1 B1 → (A2 B2 + A1 B1 →		+	+
x3 x3			
„ X A1 B1 → (A2 B2 + A1 B1 →		−	+
x4 x4 x4			
.			
.			
.			
.			
„ X A1 B1 → (A2 B2 + A1 B1 →		+	+
x9			

lation, and mated with 4 tester strains. The constitution of each of the 2 types of nuclei with respect to mating-type was deduced from the reaction with 4 testers, A1 B1, A1 B2, A2 B1, and A2 B2 (tab. 2).

The presence of auxotrophic mutations in the recombinant nucleus was determined by heterokaryotic allelic tests by the method shown in tab. 3. Each dikaryon (which contained a recombinant type nucleus $A2\ B2$ and a nonrecombinant nucleus $A1\ B1$ from the original homokaryon of the mating) was mated with a series of 9 $A1\ B1$ testers, each of which possessed a different auxotrophic mutation (the first tester listed contained $x1$, the second $x2$, etc.). The $A2\ B2$ nuclei (i.e., the recombinant nuclei) were allowed to dikaryotize the $A1\ B1$ tester strains to form 9 new dikaryons, one with each tester. The 9 new dikaryons were plated on minimal and complete media. Auxotrophy of the dikaryon $A2\ B2 + A1\ B1\ x1$, for example, indicated the homozygous condition for the $x1$ locus, and, therefore, the presence of $x1$ in the $A2\ B2$ nuclei. Growth of the dikaryon $A2\ B2 + A1\ B1\ x2$ (see tab. 3) on minimal media indicated the heteroallelic condition for $x2$ and, therefore, the absence of the $x2$ mutation in the $A2\ B2$ nuclei.

Matings were made with cultures which possessed numerous auxotrophic mutations, some of which were linked and some of which were in different linkage groups (tab. 4).

Table 4. *The types of recombination of auxotophic mutations in noncompatible dikaryotic-homokaryotic matings in which recombination of mating-type factors had occurred*

DIKARYON O—	$A1$	$B2$	$x2$ $x4$	$x5$ $x6$ $x8$			
O—	$A2$	$B1$	$x3$			$x9$	

$$\times$$

HOMOKARYON O—	$A1\ x1$	$B1$		$x7$	

$$\downarrow$$

O—	$A2$	$B2$	$x2\ x3$	$x5$ $x6$	$x9$	
CLASS I O—	$A2$	$B2$	$x3\ x4$		$x9$	
O—	$A2$	$B2$	$x2$ $x4$	$x5$ $x8$		

CLASS II O—	$A2\ x1$	$B2$		$x7$	

Two types of recombinant nuclei were recovered: 1) those which possessed some combination of auxotrophic mutations from the dikaryon (class I), and (2), those which possessed, aside from mating-type factors from the dikaryon, only auxotrophic mutations from the homokaryon (class II).

An analysis of a population of class I individuals revealed a large amount of reassociation of both linked and unlinked genes. Linked genes reassociated frequently in spite of the fact no selection was exerted for recombinants involving specific combinations of genes. The first class I individual in table 4 was the result of 2 crossovers not involving the A and B linkage groups. Several recombinations resulted from at least 2 crossovers in one linkage group, such as the third example given in tab. 4, one in each of 2 regions between $x5$ and $x6$ and between $x6$ and $x8$. In fact, the percentages of recombination of linked genes were essentially identical to those obtained via the standard sexual cycle. In one mating, which involved 8 loci in 4 linkage groups other than those involving the incompatibility factors, 43 genotypes were represented among a total sample of 61 recombinant nuclei of independent origin (ELLINGBOE, 1964).

Not all recombinant nuclei which possessed some combination of auxotrophic mutations from the original dikaryon could have had their origin in events involving only the 2 nuclei of the dikaryon. The participation of the homokaryon in recombinational events was also strongly implicated. In a dikaryotic-homokaryotic mating in which the dikaryon was homoallelic for nic-2 (nicotinic acid-requiring) and the homokaryon was nic-2+, the occurence of the wild allele in 7 out of 61 otherwise class I progeny strongly suggested its origin in recombinant progeny to be in a recombinational event rather than mutation nic-2 $\longrightarrow nic$-2+.

An explanation for the origin of the class II's is more difficult. Only the incompatibility factors appear to have been transferred. Genes in the same linkage group were not transferred. This was strikingly demonstrated in experiments in which alleles at both loci of the A-factor were transferred without transferring an auxotrophic mutation located between the loci (ELLINGBOE, 1963). The involvement of all 3 nuclear types present in the dikaryotic-homokaryotic mating is clear, but the means of effecting transfer of specific factors is not.

The specificity of the transfer process has led to the development of an episome postulate. Genetically, via the standard sexual cycle, the incompatibility factors behave entirely as "normal" genes, but under special conditions, such as in the noncompatible dikaryotic-homokaryotic matings (see tab. 4), they do not appear to be transferred as "normal" genes (auxotrophic mutations, etc.). The transfer system appears to be analogus in many respects to the transfer of temperate phage, except that the transfer is probably intracellular. Specific conditions (the incompatible reaction?) may be needed to effect transfer or release of genetic material from 2 nuclei of the dikaryon. The incorporation of the genetic material into the nucleus of the homokaryon could assume a specificity of incorporation analogous to the incorporation of a temperate phage into a bacterium and the prophage's association with a specific site on a bacterial chromosome. The incompatibility factors, like a prophage, would be expected to behave as any other genetic marker in a cross via the standard sexual cycle.

Studies on somatic recombination have suggested the incompatibility factors to be functionally different entities from genes controlling primary structure of proteins involved in metabolism. Furthermore, if special conditions are needed for "Specific Factor Transfer", are the conditions established following anastomosis of noncompatible dikaryotic and homo-karyotic hyphae, or are the incompatible systems constitutive and act immediately upon hyphal anastomosis? Preliminary cytological obser-vations suggest the former to be correct. There are no cytologically obser-vable effects immediately following anastomosis of hyphae, but a higher frequency of abnormal hyphal tips adjacent to anastomoses has been observ-ed in noncompatible than in compatible dikaryotic-homokaryotic matings (SICARI and ELLINGBOE, unpublished).

D. Summary

At least 3 different mechanisms for genetic recombination other than heterokaryosis have been demonstrated in vegetative cells of the basidio-mycetes. The mechanism in the Ustilaginales involves mitotic crossing over analogous to that described in *Drosophila melanogaster* and the para-sexual cycle described in *Aspergillus nidulans*. Whether the process of haploi-dization in *Ustilago maydis* is a mitotic or meiotic event is not clear.

In the Hymenomycetes, at least 2 mechanisms of recombination have been demonstrated. In one mechanism, the process of haploidization is indistinguishable from meiosis in that linked genes reassociate as frequently in somatic events as in sexual events. A second mechanism, termed "Speci-fic Factor Transfer", was found which reassociated only the incompatibility factors. Genes linked to the incompatibility factors were not transferred. The demonstration of transfer of incompatibility factor specificities without transferring linked, nonselected, auxotrophic mutations, even when be-tween the 2 loci of the A-factor, strongly suggests a uniqueness of the factors governing sexual incompatibility in fungi.

References

CHRISTENSEN, J. J.: Studies on the genetics of *Ustilago zeae*. Phytopathol. Z. **4**, 129—188 (1931).

CROWE, L. K.: The exchange of genes between nuclei of a dikaryon. Heredity **15**, 397—405 (1960).

ELLINGBOE, A. H.: Somatic recombination in *Puccinia graminis tritici*. Phyto-pathology **51**, 13—15 (1961).

— Illegitimacy and Specific Factor Transfer in *Schizophyllum commune*. Proc. Natl. Acad. Sci. (U.S.) **49**, 286—292 (1963).

— Somatic recombination in Dikaryon *K* of *Schizophyllum commune*. Genetics **49**, 247—251 (1964).

ELLINGBOE, A. H. and J. R. RAPER: Somatic recombination in *Schizophyllum commune*. Genetics **47**, 85—98 (1962).

HOLLIDAY, R.: Induced mitotic crossing-over in *Ustilago maydis*. Genet. Res. **2**, 231—248 (1961).

NELSON, R. R., R. D. WILCOXSON, and J. J. CHRISTENSEN: Heterocaryosis as a basis for variation in *Puccinia graminis* var. *tritici*. Phytopathology **45**, 639—643 (1955).

PAPAZIAN, H. P.: Exchange of incompatibility factors between nuclei of a dikaryon. Science **119**, 691—693 (1954).

PARAG, Y.: Studies on somatic recombination in dikaryons of *Schizophyllum commune*. Heredity **17**, 305—318 (1962).

PRUD'HOMME, N.: Recombinaisons chromosomiques extra-basidiales chez un basidiomycete "*Coprinus radiatus*". Annales de Genetique **4**, 63—66 (1963).

PONTECORVO, G.: The parasexual cycle in fungi. Ann. Rev. Microbiol. **10**, 393—400 (1956).

QUINTANILHA, A.: Etude Genetique du phenomene de Buller. Bol. Soc. Brot. **13**, 425—486 (1939).

RAPER, J. R., and M. OETTINGER: Anomalous segregation of incompatibility factors in *Schizophyllum commune*. Revista Biol. **3**, 205—221 (1962).

ROWELL, J. B.: Segregation of sex factors in a diploid line of *Ustilago zeae* induced by alpha radiation. Science **121**, 304—306 (1955).

STERN, C.: Somatic crossing-over and segregation in *Drosophila melanogaster*. Genetics **21**, 625—730 (1936).

SWIEZYNISKI, K. M.: Somatic recombination of two linkage groups in *Coprinus lagopus*. Genetica Polonica **4**, 21—36 (1963).

WATSON, I. A.: Further studies on the production of new races from mixtures of races of *Puccinia graminis* var. *tritici* on wheat seedlings. Phytopathology **47**, 510—512 (1957).

WATSON, I. A., and N. H. LUIG: Somatic hybridization in *Puccinia graminis* var. *tritici*. Proc. Linnean Soc. N. S. Wales **83**, 190—195 (1958).

Discussion

Mather: Are there cases where only one of the two subunits is transferred?

Ellingboe: Both subunits of A as well as the B factor were transferred in our experiments. It's difficult to get conclusive evidence about transfer of only one of the subunits or of only one of the factors because such individuals would have a parental combination of markers from the nuclei of the dikaryon, and such individuals could also be obtained through meiotic recombination.

We have observed a higher percentage of parental combination of markers than expected among the class I recombinants. This high frequency suggests that we *might* have had transfer of only one of the two factors rather than a meiotic recombination.

Elliot: Have the centromeres been located?

Ellingboe: Yes. For the A chromosome. Our primary evidence that only the incompatibility factors are transferred has come from the experiments in which the two subunits were transferred without the simultaneous transfer of the *pab* marker, which is between $A\alpha$ and $A\beta$, or the *x-15* marker, which is distal to A.

Mather: Has this phenomenon occurred to any biochemical markers?

Ellingboe: I wouldn't have detected it, because recombinational events were detected on the basis of recombination of the incompatibility factors. Another experimental method might detect this.

Snider: Is there any evidence from your results that the mating-type factors are different from structural genes in their gene action?

Ellingboe: We have no evidence as to their function. We only know that the factors are unique from the standpoint of transfer in dikaryotic-homokaryotic matings.

Mather: There are many things which we understand at the genetic level, without gaining any real knowledge of the way they function.

Parag: In connection with the episome hypothesis, one of our former students, working under the assumption that such episomes should be more sensitive to acridine dyes as mutagens, treated a very small sample of a mixture of macerated mycelia of *A*1 *B*1 and A2 B1, and he detected several dikaryotic fruit bodies. In these fruit bodies, he found at least one *B* mutation and one *A* mutation. If these results are significant, they could give some indirect support to your episome hypothesis.

Somatic recombination in the Basidiomycete Coprinus radiatus

by

NICOLE PRUD'HOMME

The sexual behavior of *Coprinus radiatus* is determined by tetrapolar incompatibility. The mating competence of homokaryotic strains depends on two series of *A* and *B*-factors (for more details, see the paper of DAY p. 31). Additional genes are known, however, that prevent the formation of a dikaryon between homokaryons carrying different *A*- and *B*-factors. Thus, if the three following strains are confronted:

1. *A*1 *B*1 +
2. *A*2 *B*1 n
3. *A*1 *B*2 n,

(where *n* is a gene leading to a dwarf phenotype) none of the three combinations of two can give a dikaryon. These three strains, however, are able to form a triheterokaryon, i.e. a mycelium with the three nuclear types.

The phenotype of this triheterokaryon differs according to the strains, which are used. They are always intermediate in their characters between those of wild-type homokaryons and dikaryons, but their growth is weak and inferior to that of a dikaryon. From time to time, sectors of distinctly dikaryotic phenotype and rapid growth appear. The study of the genotypic constitution of these dikaryotic sectors is the subject of this paper. The linkage maps of the genes which will be used have been established by PRÉVOST (1962).

The nuclear constitution of the dikaryotic sectors has been determined by two methods:

1) Analysis of progeny of fruitbodies, either by the examination of spore tetrads or of random spores. This method permits a determination of the nature and number of the alleles at marked loci. It is the only way by which the presence of polysomy can be unambiguously demonstrated.

2) De-dicaryotisation, by macerating in a waring blender, mycelia that come from the dikaryotic sectors. This method permits the observation of the phenotypes of the different haploid constituents of the sectors, and hence, the association of markers in the two nuclear types which exist at the base of the subsequently formed fruitbodies. If there are more than two nuclear types in a sector, it should be possible to demonstrate them.

By using these two methods, we found that the "dikaryotic-like" sectors had one of the initial nuclear types and one or more new type, compatible with the first one and originating from an exchange of genetic material between the two initial types. Usually, these exchanges occur between nuclei of compatible types, e.g., between $(A2B1 \text{ n})$ and $(A1 B2 \text{ n})$. $(A1 B1 +)$ nuclei select recombinations of A and B which are compatible with them, i.e. $(A2 B2 \text{ n})$. More rarely, exchanges occur between nuclei of incompatible mating types, such as between $(A1 B1 +)$ and $(A1 B2 \text{ n})$ or between $(A1 B1 +)$ and $(A2 B1 \text{ n})$. *In the triheterokaryons, therefore, somatic recombinations occur.*

We have performed several experiments using a wide range of genes as markers. Through the study of their segregations in the new nuclei, we are able to offer suggestions for the mechanisms which produce these extra-basidial exchanges of genetic material.

For example, the constitutions of three strains which were used to form triheterokaryons in a series of experiments were:

Chromosome	III	I	II	VI	V	IV
1.	$A2$	$B1$	$+$ $ad28$	$ur5$	$+$	$+$
2.	$A1$	$B1$	col $ad28$	$+$	$arg10$	$+$
3.	$A2$	$B2$	col $+$	$ur5$	$+$	$arg8$

col, a morphological dwarf mutant and *ad28* (adenineless) are located on the two arms of Chromosome II at a distance of 38 cross over units.

The genes used in the selection of the recombinants were the incompatibility factors and *col*. In recombinant nuclei, associations of other gene markers are not generally selected, since the triheterokaryons were grown on complete medium.

Nevertheless, an exception was selection in favour of $ad28^-$ and col^+, when the recombinants were produced between nuclei of types 1 and 2, or between 1 and 3.

Analysis of 250 sectors of independent origin, obtained in our experiments using the same markers as in the example above in various combinations, or other combinations of markers, especially genes linked either to A, to B, or to each other, led to the following results:

1) In the new haploid nuclei, markers independant of A and B and independant of each other can be re-associated. Thus in the experiment cited above for example, the following nuclear constitution can be obtained:

$A1 B2$ *col ad28 ur5 arg10 arg8*

2) Genes located on the same chromosome, even when far apart, do not recombine, or, if they do, it is with a frequency far lower than the frequency of meiotic recombination. For example, in one experiment, in which the

genes used for selecting recombinants were *A*, *B* and *on* (a morphological mutation linked to B) and where the markers were *col*, *ad28*, *ur5* and *arg10*, we found 21 new nuclei recombined for *ur5* and *arg10*, situated on different chromosomes, and no new nucleus recombinant for *ad28* and *col*, situated in Chromosome II 38 c.o. units apart.

3) Among the newly constituted nuclei, we have been able to show, by analysis of the progeny of fructifications developed from sectors, that many aneuploids can be found; some are simple disomics, others are multiple disomics. All of the kinds of aneuploids to be expected with the markers used were observed.

4) Sometimes de-dikaryotisation permitted us to detect several new nuclear types per sector.

We can easily explain these results if we suppose that the new nuclei appearing in the triheterokaryons originally come from the diploidization through fusion of two haploid nuclei, followed by progressive haploidisation through a mechanism analogous to that described in the Ascomycetes (PONTECORVO, 1956). This hypothesis explains the recombinations of markers independant of *A* and *B*, the presence of several new types in a single sector, and the high frequency of aneuploids. The apparent total linkage of markers situated on the same chromosome excludes the possibility of extra-basidial meiosis.

There can be no doubt that diploids are produced. We have been able to isolate two diploid strains by de-dikaryotisation of two independant sectors. One is

$$A1/A2 \; B1 \; +/B1 \; on \; ad28/+ \; ur5/ur5 \; arg10/+$$

and has identical *B*-factors. It is very stable. The other is

$$A1/A2 \; B1 \; on/ \; B2 \; on \; col \; ad28/+ \; + \; ur5/+ \; arg10/ \; +.$$

It is relatively stable, enough to be maintained, but it can give a variety of aneuploid and haploid nuclei.

We have observed that, by the maceration of triheterokaryons in a waring blender, diploid strains of this latter type are very easily isolated. As soon as the triheterokaryons are formed, diploid nuclei appear with high frequency.

The new nuclei detected in the dikaryotic-like sectors originate, at least in some cases, from diploid nuclei.

One fact, however, requires emphasis. In all the experiments, we find a significant and constant excess of new nuclei showing a parental association of the independant markers which have not been used to select for nuclear recombinations.

If, in an experiment, two chromosomes are marked by *A* and *B* and two others by the genes *x* and *y*, and if we select recombinants for *A* and *B*, the parental associations for *x* and *y* are about four times more frequent than the recombinant associations. But, if six chromosomes are marked and if we select not only recombinants for *A* and *B* but also recombinants for two other independant markers, the parental and recombinant associations of *x* and *y* have the same frequency (tab. 1).

This means that there are two types of recombination:

1) One leads to a random association of markers independant of A and B and independant of one another.

2) The other most often leads to a parental association of the same markers.

These two types of recombination might correspond to two mechanisms. Some new nuclei could come from the "transfer" of a single chromosome, for example, at the moment of conjugate division, to correspond to the second type of recombination.

Table 1. *Scheme to show the frequency of recombination of 2 genes located on different chromosomes*

Nuclear constitution of 2 parental types	genes selected for	Ratio of	
		Parental types	Recombinant types
	$A1\ B2$	$(x\ +)\ (+\ y)$ 4	$(x\ y)\ (+\ +)$ 1
$A1\ B1\ m + x +$ $A2\ B2 + n + y$			
	$A1\ B2\ m\ n$	1	1

We might also suppose that all the new nuclei could come from diploid nuclei and that, when these become haploidized, the parental associations of markers are strongly selected in certain cases. We know, from the study of haploidisation of diploid strains, that such selection cannot occur after these strains have already been isolated. We must notice, however, that only a few diploid nuclei are formed in a mycelium of which the majority nuclei are of the original types. When diploid nuclei revert precociously to the haploid state, it is possible that those nuclei are strongly selected which most closely resemble the genotype of the surrounding parental nuclei.

References

PONTECORVO, G.: The parasexual cycle in fungi. Ann. Rev. Microbiol. **10**, 393—400 (1956).

PRÉVOST, G.: Etude génétique d'un Basidiomycète: Coprinus radiatus, FR. ex. BOLT. Thèse de Doctorat Paris (1962).

Discussion

Casselton: When you have a dikaryon with one haploid and one diploid component, how stable is the diploid nucleus in the dikaryon?

Prud'Homme: That depends on the diploid strain. Diploid nuclei which have two like B's are very stable. Those which have two different A's and two different B's are more unstable. In the latter case, the segregations observed in the progeny of fruit bodies arising from haploid-diploid dikaryons are at times triploid segregations, but more often aneuploid or diploid segregations.

Casselton: We find, in *Coprinus lagopus*, that our diploids are very stable as homokaryons but unstable when combined in a dikaryon with a haploid nucleus. We haven't, though, put two diploid nuclei in a dikaryon.

Parag: Did you look for somatic recombination in the dikaryons in the absence of di-mon matings?

Prud'Homme: Yes, I have tried that, but the recombinants which I thought I would be able to select for were not observed. Nevertheless, recombinants are produced, and I obtained them purely by chance.

Burnett: Have you formed dikaryons from a haploid and a diploid strain?

Prud'Homme: Yes, and cytological observations have been made on them. Mr. MOTTA in our laboratory has found that the two nuclei which are present in each cell are of different sizes, even though the nuclear pairs in normal dikaryons are to all intents and purposes equal in size.

Snider: Do you have any evidence for specific transfer of incompatibility factors such as was reported by Dr. ELLINGBOE?

Prud'Homme: No. I haven't observed recombinations between the incompatibility factors and the markers which are linked to them.

Snider: The experiments you describe involved triple confrontations between homokaryons. Have you observed, in any unreported experiments, indications of nuclear migration into already formed dikaryotic mycelia?

Prud'Homme: No, but we have not looked for this.

Incompatibility and nuclear migration

by

PHILIP J. SNIDER

Nuclear migration in fungi is closely associated with both heterokaryosis and sexual reproduction. Formation of heterokaryons can be enhanced by nuclear migration; the control of breeding can restrict migration. As this symposium bears witness, much effort is now aimed toward understanding how incompatibility genes regulate migration, heterokaryon formation, and breeding. The investigation of nuclear migration has developed rather slowly since its initiation by BULLER (1931), and although substantial progress has been made, much obviously remains to be done.

For a variety of fungi, marked nuclei introduced at one point in a mycelium can be recovered sometime later far removed from the point of entry. While this sort of displacement may occur in several ways, in at least some instances (BULLER, 1931; SNIDER, 1963) it has been demonstrated conclusively to be nuclear migration, the transfer of nuclei from one strain into and throughout the pre-grown mycelium of another. Migration between strains can be either reciprocal (bilateral) or one way (unilateral). The terms *donor* and *recipient* will serve to specify the origin and polarity of migration. Unless noted otherwise, previously unpublished results presented here refer to *Schizophyllum commune*.

That nuclear migration is regulated became apparent about the same time migration was discovered. In *Neurospora*, nuclei that entered the protoperithecium during outcrossing failed to migrate into the vegetative hyphae but were recovered in ascospores (DODGE, 1935). A similar restriction upon migration evidently applies to mycelial pairings of $A \times a$ strains, even though heterokaryons constituted of strains of opposite mating-types can grow if forced (SANSOME, 1946). In Basidiomycetes, correlation of the B incompatibility factor with control of nuclear migration was observed almost simultaneously in two species (FULTON, 1950; PAPAZIAN, 1950). While PAPAZIAN did not interpret his observations this way, FULTON (1950, footnote 9) did.

The biological significance of nuclear migration is not very obvious. Since heterokaryons can form by growth, with or without nuclear migration, what, if any, is its selective advantage? A likely possibility is that migration increases the rate of heterokaryon-formation, as rate may be an especially crucial factor in the early stages of mycelial development. In any pairing of homokaryotic strains, heterokaryon-formation and homokaryotic overgrowth may be regarded as apposing tendencies. If heterokaryosis must be extended by growth alone, it can only do so from a narrow stricture at the limits of the contact between the mycelia. The chances seem to be against extension of heterokaryosis at the expense of homokaryosis, unless the heterokaryotic growth is substantially faster than that of the homokaryons. Development by growth *and* nuclear migration could enable heterokaryosis to begin at once from the entire mycelial front of paired homokaryons. In this instance, it is only necessary to assume that nuclear migration would be substantially faster than hyphal tip growth to conceive how nuclear migration alone might favor heterokaryon-formation over continued homokaryosis or even one type of heterokaryosis over another.

A. Methodology

The requirements for detecting nuclear migration and for estimating its rate are not necessarily identical.

I. Detection of migration

Visual observation of nuclear migration is possible in theory, but to see living nuclei is difficult and to distinguish between donor and recipient nuclei presents many problems that have not yet been satisfactorily resolved. Nuclei are sometimes found straddling the septal pore in fixed material viewed in a light (MACDONALD, 1949; SANFORD and SKOROPAD, 1955) or electron microscope (SHATKIN and TATUM, 1959), but the relevance of such observations to active nuclear migration is uncertain.

Genetic evidence has provided the classic and most reliable means of detecting migration. Clamp connections (BULLER, 1931) and morphological mutants (SNIDER, 1963) have been employed successfully. A morphological mutant of *Schizophyllum*, *puff*, is uniquely favorable material. When a *puff* strain is used as a recipient, invasion by wild-type donor nuclei converts *puff* hyphae individually into normal dikaryotic hyphae (fig. 1). A piece of the converted *puff* hypha is then transferred to fruiting medium. Several genetic markers besides *puff* are included in the donor and recipient strains, and normal meiosis confirms that whole nuclei from the donor migrated within preformed hyphae of the recipient strain. The morphological distinction between donor and recipient hyphae avoids any confusion between intrusive growth and nuclear migration.

Kinetic evidence has been the most common means of demonstrating nuclear migration (Buller, 1931; Dowding and Buller, 1940; Fulton, 1950; Snider and Raper, 1958; and Swiezynski and Day, 1960b). The

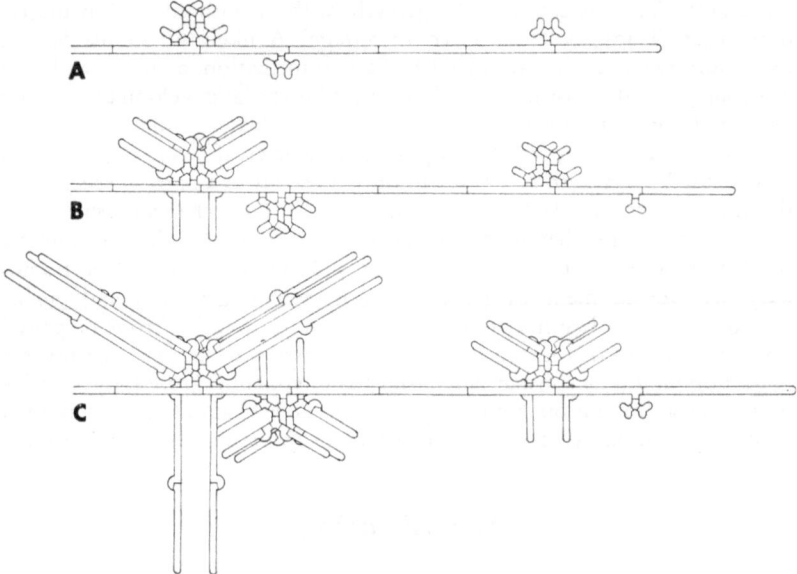

Figure 1. Diagrammatic representation of the *puff* technique. The change in time (*A—C*) of mutant morphology (*A*) to wild-type growth indicates the occurrence and direction of nuclear migration (left to right here) from a *p*⁺ donor strain (not shown) into a single hypha of the *puff* recipient strain

special advantage of this method is that it may be done in thickly grown cultures. The presence of recipient nuclei only or of donor and recipient nuclei is distinguished in material removed from the experimental culture and grown in isolation (Plate 1, b). Either morphological indicators or nutritional deficiencies and selective growth may serve for this purpose. A clean separation of migration from intrusive growth, however, is often difficult.

The crux of the problem is that the evidence for migration is indirect. A recipient strain is grown in one culture, and then a donor strain of another genotype is sumultaneously implanted on the recipient mycelium and inoculated separately in sterile medium. Nuclear transport in the recipient mycelium is then compared with the linear growth of the donor strain in separate culture; any intrusive growth by the donor strain is assumed to be neither faster nor farther than its growth in isolation. Nuclear transport in the paired culture several times farther than growth of the donor strain in isolation is accepted customarily as evidence for migration.

Autoradiography offers another possible method of detecting nuclear migration, but the author is not aware of any successful development of it.

The non-visual techniques for detecting migration assume that any small number of donor nuclei will be detected in a test sample from the recipient mycelium. This may not be so. Will a single donor nucleus, for example, be invariably detected in a test plug 2 mm in diameter, or is a threshold number of nuclei required? Although no direct test has come to mind in the past 8 years, there are several inconclusive observations consistent with a threshold number of nuclei. The time required for test plugs to respond on the selective medium increases with migratory distance, as though the response-rate were related to the initial number of donor nuclei in each test plug. Most of the reliable results with *S. commune* have been obtained with recipient strains that give a slight amount of background growth on the selective medium. (Recipients of non-leaky mutants give a little background growth when a bit of specific supplement is carried over in the test plug.) Enhanced sensitivity is expected if detection were dependent upon background growth increasing the donor nuclei to a threshold number. The possibility of undetected migration evidently cannot be dismissed lightly.

Detection is also affected by the type of test sample. It has been clearly shown that donor nuclei are detectable centimeters farther from the implant in test plugs (2 mm in diameter; $c. 10^4$ cells/plug) than in hyphal fragments (1-10 cells/fragment) taken simultaneously from identical parts of the recipient culture (ELLINGBOE, 1964). While fragments may be useful for estimating saturation of the recipient mycelium with donor nuclei, fragments are surely inappropriate for detecting migration. Their low sensitivity probably results from a substantial loss of cells during maceration in addition to a possible threshold effect.

II. Estimating the rate of nuclear migration

The determination of the rate of nuclear migration has in no case been determined by direct observation. The rate of general protoplasmic streaming, however, has been measured in several Ascomycetes, and these rates were interpreted as indicative of the rates of nuclear migration (DOWDING, 1958; DOWDING and BAKERSPIGEL, 1954).

The more reliable estimates of nuclear migration in Basidiomycetes have been made by indirect methods, the most reliable of which is based upon the rate at which the front of a migratory population sweeps across a recipient mycelium. While the mere detection of migration is of secondary interest here, its overall efficiency must be higher than the low frequency of recovery that may suffice to demonstrate migration. Reliable data might be obtainable from *individual hyphae*. Use of this variation has not been noted, although it is feasible with the *puff* mutant. If the lag between the arrival of a migrant and the local expression of its phenotype is large and variable, however, so will be the error in the rate-estimate.

The possible disadvantages of phenotypic lag can be bypassed if rate-estimates are based upon the motion of donor population fronts. Small, cleanly cut pieces of recipient mycelium, transferred to separate cultures, fix the location of donor nuclei with time. To detect the donor nuclei, either by use of morphological or nutritional markers, the period of incubation can be extended in proportion to phenotypic lag without effect upon the estimate of migratory rate.

Precise definition of the migratory front at several times is crucial. This is a separate matter from merely detecting the presence of donor nuclei or ensuring that transport is by migration. Acceptance of a front as well defined is probably safe if no gaps appear in the expanding pattern of positive test samples. The shape of the front seems predictable in some experimental systems, and this can serve as an additional check on the efficiency of detection. For 10- and 15-cm recipient cultures spot implanted at the center, the front has been shown (within 5 mm) to move as an expanding circle, and donor nuclei were detected in all the test plugs between the front and the implant (SNIDER and RAPER, 1958, fig. 5). In a rectangular-shaped recipient, with a strip implant across one short side, the front is probably a straight line at right angles to the direction of migration (EL-LINGBOE, 1964). The apparent advantages of rectangular geometry should be further explored. Since a low population density of donor nuclei, which may often characterize most of the migratory period, can be masked by the high sensitivity of large test plugs, only data from sharply defined fronts should be accepted for estimating rate.

Ordinary cultures, grown from a point inoculum at the center of a Petri plate and thus termed *radiate* recipients, are not recommended for estimating the rate of migration. The geometry of their hyphal matrix, their radially oriented age-gradient, and their lack of definable migratory fronts are altogether disadvantageous for quantitative experiments. *Reticulate* recipients, established by inoculation with a suspension of mycelial fragments, are much better for quantitative work than radiate recipients (SNIDER and RAPER, 1958). Reticulate recipients have no dominating age-gradient or preferred orientation of hyphae. The rate of migration is somewhat slowed, the density of the donor population is sustained at detectable levels, and the migratory fronts are definable and symmetrical.

B. Results and discussion

I. Facts about nuclear migration

The two most intriguing aspects of nuclear migration are the detailed mechanism underlying the rapid and oriented migration of nuclei over long distances and the genetic regulation of migration. Although there is a considerable body of facts relating to the process of nuclear migration, there is even now insufficient information at the cellular and biochemical level to permit any satisfactory resolution of either of these problems. Any attempt to resolve these problems or to formulate reasonable working hypotheses must accommodate the following established features of the migratory process.

Rapid nuclear migration has been demonstrated by indirect means in certain Ascomycetes (DOWDING and BULLER, 1940; DOWDING and BAKER-SPIGEL, 1954), and comparable rates of intercellular cytoplasmic flow along individual hyphae have been visually observed in the same organisms (DOWDING, 1958; SNIDER and G. N. BISTIS, unpublished). Generalized protoplasmic streaming has a definite overall pattern in undisturbed homo-karyotic mycelia of *Gelasinospora*, *Neurospora*, and *Ascobolus*. The streaming is polarized from the center of the mycelium toward its growing circumference. The most rapid streaming is limited to a few, optically dense, major hyphae in the older part of the mycelium, while most of the hyphae in this part are highly vacuolated, without conspicuous streaming. As it nears the growing fringe, the streaming is diverted into numerous branch hyphae and is slowed. An exact correspondence is noted between tips measurably elongating and visible streaming to the tips. Although the relationship between generalized cytoplasmic streaming and the transport of nuclei has not been established, a direct correlation appears quite plausible in these Ascomycetes.

Far more intensive study, however, has been devoted to nuclear migration in Basidiomycetes, which differ from the Ascomycetes rather strikingly in detailed hyphal and cellular morphology and in which there are no reported observations of generalized cytoplasmic streaming with which nuclear migration has been correlated. BRACKER and BUTLER (1964), however, report generalized streaming in *Rhizoctonia (Pellicularia) solani* comparable to that observed in the Ascomycetes. The following facts about nuclear migration have been established for members of the Basidiomycetes.

Previously unreported experiments have revealed nuclear exchange between mycelia to be an infrequent event. The frequency of nuclear transfer into a recipient mycelium was determined along 2-meter confrontation lines between compatible mycelia. At 32° C, 3 hours after implantation, 10 transfers had occurred; by 12 hours, the number had increased to 200. Opposed hyphal tips of either mate in the 2-meter line of confrontation numbered approximately 10^5, and the frequency of transfer per hyphal pair ranged from 3×10^{-5} to 2×10^{-3}. The experiment was not designed to test whether the exchange at any point of fusion is unilateral or reciprocal.

The rate of nuclear migration is several times the rate of hyphal-tip growth. This is perhaps the widest based generality about migration in filamentous fungi (tab. 1). The kinetic method for detecting migration and the selective advantage suggested for migration depend upon this differential. The modest rate of 1 mm/hr carries a donor nucleus through about 10 recipient cells each hour; under the same favorable conditions of vegetative growth, nuclei in apical cells will divide no more than once an hour (SNIDER and RAPER, 1958). These facts would appear to provide a practical answer to BULLER's query of whether the migrating nucleus divides in each transversed cell.

toward and through the older, central portions of the culture (Buller, 1931).

The nuclei might be capable of self-propulsion. The nuclear envelope may well be associated with submicroscopic structures, such as contractile elements, microtubules, etc., which, in the viscous milieu of the cytoplasm could impart to the large nucleus an oriented movement. Such a hypothesis of active motility has no direct evidence to support it, but a few observations of intracellular and limited intercellular nuclear movement in fungal cells, in the Basidiomycetes especially, suggest autonomous movement of the nuclei through distances that are quite extended relative to nuclear size.

The possibility of electrical charges being responsible for nuclear movement appears remote, since the potential differences required would be very high.

III. Experiments on regulated migration

The investigation of regulated nuclear migration is mainly in the descriptive phase. Prime attention is being given to the search for novel examples and to the check of new discoveries. The aspects of migration possibly subject to specific regulation include the frequency of initiation, the direction, rate, pattern and extent of migration and the population density of donor nuclei. In initiation from heterokaryotic donors, nuclear selection offers an additional aspect. An involvement of incompatibility genes is evident in most of the examples to be given, but the specific aspect of migration affected is not yet known in every instance. After the evidence for regulation is considered for some of the best established examples, current progress in investigating the modes of action on several of the aspects affected will be discussed.

Unilateral mating (Raper, 1953), or unilateral "diploidization" (Brodie, 1948), has been reported for a number of fungi. The direction of nuclear exchange and migration is the aspect regulated, as exchange and migration take place in one direction only between paired, dissimilar strains. Two basic types of unilateral migration may be recognized. In one type, a genetic basis for the regulation has been attributed to morphological mutations at any of various loci (Raper and Miles, 1958). The mutant strain acts only as a donor in pairings between mutant and wild-type homokaryotic strains. In the second type, wild-type dikaryotic mycelia act only as donors.

Unilateral donors are uniquely suited for demonstrating the extent of intrusive growth expected in experiments concerned with kinetic evidence for nuclear migration. A *uracilless streak* homokaryon (*streak* is a unilateral donor morph-mutant) was combined in a compatible pairing with an *arginineless* homokaryon: *A1 B1 ura-1 arg-2+ st $\rightleftarrows$ A2B2 ura-1+ arg-2 st+* (each arrow indicates a direction tested for transport of donor nuclei). In one set of cultures the *streak* strain was the reticulate "recipient"; in another set of cultures the morphologically wild-type strain was the reticulate recipient. Since *streak* acts only as a donor (Papazian, 1951; Raper and Miles, 1958), recovery of donor nuclei from the *streak* "recipient" must indicate the extent of intrusive growth, not nuclear migration; the morphologically wild-type recipient serves as a control for normal migration in the opposite direction and for the formation of hyphal fusions. The general form of the experiment, with central implants and test plugs cut on 4 radii, was as illustrated in Plate I, a and b, except that 10-cm Petri plates were used and only 8 evenly spaced test plugs were taken from each radius. The period of incubation at 30° C, the time from implantation to removal of the single set of test plugs, would have allowed unrestricted nuclear migration for about 470 mm (10 times the distance from implant to the edge of the plate) and unrestricted intrusive growth for 30 millimeters. From the 13 "recipient" cultures of *streak* mycelium, donor nuclei were recovered by specific selection in only 4 of the 416 test plugs. All 4 were from the locations closest to the implants and indicate clearly that intrusive growth into *streak* was no more than 5 mm in 156 hours. In the control, migrant nuclei were recovered in all 288 test plugs from the 9 recipient cultures of the morphologically wild-type recipient. Practically identical results were obtained in other experiments with dikaryotic unilateral donors (see below). The data from the 2 unilateral donor systems essentially rule out intrusive growth as a possible explanation for any significant nuclear transport in reticulate recipients of *S. commune*.

Nuclear migration in Basidiomycetes is regulated selectively by one of the two incompatibility factors (usually termed the *B* factor) in several tetrapolar species, as reported for *Cyathus stercoreus* (FULTON, 1950), *Schizophyllum commune* (PARAG and RAPER, 1960; also see FULTON's 1950 reference to PAPAZIAN, 1950) (Plate I, c—f), *Coprinus lagopus* SWIEZYNSKI and DAY, 1960a, b), and at least 8 other species (TAKEMARU, 1961). The remaining examples of regulation will include those observed in pairings between homokaryons, between a homokaryon and heterokaryon, and between heterokaryons.

Any pairing between homokaryons with *B*= factors is reported to show no migration, but the meaning of "no migration" here should be weighed with more than usual caution. *Radiate* recipients (and probably non-optimal conditions for migration) were used in all the cited researches. At stake here is the firm identification of the affected aspect of migration. Is it a failure of initiation, a greatly reduced rate, or something else? Since the formation of heterokaryons is possible in all 4 incompatibility combinations (SWIEZYNSKI and DAY, 1960a; RAPER and SAN ANTONIO, 1954; PARAG and RAPER, 1960; and MIDDLETON, 1964), a failure to form hyphal fusions cannot be the answer.

An extensive attempt was made to screen for kinetic evidence of nuclear migration in $A \neq B=$ and $A= B=$ pairings of *S. commune*. The $A \neq B \neq$ and $A= B \neq$ pairings were always included as controls. Six to 10 replicate cultures were included of each of the 4 pairings in every run of the experiment. The experiment was run a total of 5 times, and only a small fraction of the data can be presented here. There were several other features common to the whole series of experiments. A variety of donor strains were separately implanted in reticulate recipients of various *arg-2* strains on migration complete medium (SNIDER and RAPER, 1958) and inoculated in isolation in sterile plates of the same medium. The *arg-2* mutant is non-leaky but gives a little background growth in plugs from this medium; the background growth is permitted deliberately, as it seems to increase the sensitivity for detecting donor nuclei (see Methodology) and is easily distinguished from growth that indicates the recovery of donors (Plate I, b). After implantation, the cultures were incubated two times longer than is needed for donor nuclei to migrate to the edge of the plates in $A \neq B \neq$ compatible controls; then, a single sample of test plugs, 8 from each of 4 radii, were transferred from each culture to media specifically selective for the donor nuclei. Various temperatures between 22 and 35° C were used, each in separate runs of the experiment, to obtain the largest possible separation between the rates of migration and intrusive growth (these experiments predated the unilateral-donor experiments).

In most instances, every plug from $A \neq B \neq$ and $A= B \neq$ pairings was positive. Equal rates of migration are not implied, however, as the experimental design was only for detecting migration. Donor nuclei in both sets of $B=$ pairings were often recovered 20 to 40 mm from the implant (Plate I, g—j). This is up to 20 times farther than any plausible intrusive growth, which the experiments with unilateral donors show to be almost nil in reticulate recipients. Even the controls of the growth of donor strains in isolation, which indicate the limit of intrusive growth if it were unrestricted, were typically no more than 1/2 the maximal distance at which donor nuclei were recovered. These results constitute strong kinetic evidence for nuclear migration in $A \neq B=$ and $A= B=$ pairings of S. commune. Under the favorable conditions of these experiments, the B factor may affect the rate of migration, the pattern of migration (which appears irregular), and possibly some other aspects of migration. The B factor, however, does not seem to prevent the initiation of migration or to restrict greatly its ultimate extent.

The rate of nuclear migration in $A \neq B \neq$ and $A= B \neq$ pairings has been reported as probably equal in S. commune (SNIDER and RAPER, 1958) but different by about a factor of 2 in C. lagopus (SWIEZYNSKI and DAY, 1960b). Since the estimates in both instances were made in radiate recipients, where it is practically impossible to define migratory fronts and maximal slopes for rate estimates, the measurements should probably be done over in reticulate recipients before any firm conclusion is made about the effect of $A=$ factors on the rate of migration. Variations characteristic of specific pairings (fig. 2) and therefore irrelevant to incompatibility gene action must also be excluded.

Among the pairings between homokaryons and various types of heterokaryons, the possibility of nuclear migration from a homokaryon into a dikaryon has previously been practically unexplored. SWIEZYNSKI and DAY (1960b) noted, without details, that migration was not observed into dikaryons of Coprinus lagopus; presumably, their trials were not extensive. Electron microscope observations of septal pore structure in dikaryons of several species led MOORE and MCALEAR (1961) to predict that dikaryons probably would not act as recipients. No experiments were mentioned.

An experiment involving two different donors was designed to answer two questions: what happens when 2 migratory fronts meet in a common recipient that is compatible with both of the donor components, and can any migration into a dikaryotic recipient be detected? The preparation of the recipient cultures and the conduct of the experiment proceeded as follows.

Macerated mycelia of *A2 B2 arg-2* and *A38 B38 arg-2* homokaryons were mixed in molten agar of migration complete medium, which was then poured into sterile, 10-cm Petri plates. When the agar solidified, circles were cut in the agar and their centers removed. The cleared centers were then poured with agar medium containing a macerate of *A2 B2 arg-2* homokaryon. Discs of an *A4 B4 ura-1* donor strain were implanted after 48 hours, the time required for growth and fusion of the recipient mycelia. Migration was thus initiated in opposite directions at about the same time. The width of the homokaryotic recipient zone was 10, 20, and 30 mm for separate treatments of the experiment. In another treatment, the homokaryotic recipient zone was omitted, so that the entire culture was formed as a single, dikaryotic recipient.

After 10 times the period needed for migration, at a rate of 3 mm/hr at 32° C, from the implant to the edge of the plate, 8 test plugs from each of 4 radii were taken from each culture and transferred to minimal plus uracil medium, which selects specifically for the *A4 B4 ura-1* donor. Every treatment and control consisted of 5 or 6 cultures.

In separate controls, migration was tested into *A2 B2 arg-2* and *A38 B38 arg-2* homokaryotic recipients. In another control, migration of *A38 B38 arg-2* nuclei toward the center of double recipients was checked in cultures not implanted with the *A4 B4 ura-1* donor. The expected donor nuclei were recovered in all the test plugs from the controls.

In the 3 double-recipient treatments, the *A4 B4 ura-1* donors migrated *half* the distance across the *A2 B2 arg-2* homokaryotic zones and then stopped (Plate I, k—n). Unrestricted migration would have passed 400 mm in the same length of time. Circular migratory fronts apparently moved centripetally and centrifugally toward one another, collided and stopped. Since all the strains in this experiment were compatible in all combinations, a principal conclusion is that freshly dikaryotized cells seem to barricade themselves immediately against the entry of a third type of nucleus. The barricade holds indefinitely. Five positive plugs were recovered from positions nearest the implant in the treatment of wholly dikaryotic recipients, but these responses are interpreted as indicating intrusive growth of not more than 5 millimeters. The *A4 B4 ura-1* donor strain grew 21 mm in isolation. No significant migration or intrusive growth was demonstrated into dikaryotic mycelia. The dikaryotic barricade seems to be a novel type of regulation because it takes place in the presence of $B \neq$ incompatibility factors. The polarity of migration seems to be the aspect affected, as dikaryotic mycelia are unilateral donors.

Migration from dikaryons into homokaryons takes place freely in the same pairing as above but with the direction of migration reversed. This is the well-known "Buller Phenomenon" and has received much attention. The most interesting possibility for regulation here is nuclear selection of one component of the dikaryotic donor over the other and of either dikaryotic component over a dikaryotic pair (CROWE, 1960; ELLINGBOE and

Raper, 1962; Crowe, 1963; and Ellingboe, 1964). The incompatibility factors have been implicated in nuclear selection, especially by Crowe (1963), who found selection to be strongly correlated with specific allelic combinations within A. Nothing definite is known, however, about the mode of selection. Although the recent results of Ellingboe (1964) fall short of identifying the mode, place or time of nuclear selection, convincing kinetic evidence was found for migration of both donors from dikaryons into homokaryotic recipients in pairings in which either or both dikaryotic nuclei have an incompatibility factor in common with the recipient homokaryon. Selection of $A\neq B\neq$ heterokaryosis over hemi- and non-compatible heterokaryosis is thus not qualitative, as previously assumed, and migration in $B=$ factor combinations is possible in multiple pairings also including $B\neq$ factors.

The other pairings between homokaryons and heterokaryons apparently have not been surveyed systematically for the occurrence of bilateral migration. Raper and San Antonio (1954) recovered a third type of nucleus (second donor) from $A=B\neq$ recipients. The results of Ellingboe (1964) suggest that migration may be possible into any heterokaryotic recipients that are not dikaryotic.

Pairings between heterokaryons seem to be the least explored of all. Swiezynski (1961b) reports an exchange of nuclei at the confrontation line, without subsequent migration, in $A\neq B\neq$ to $A\neq B\neq$ pairings.

IV. Possible mechanisms of regulation

Little is known about the mechanism by which any particular aspect of nuclear migration is regulated, and only a limited amount of factually based speculation seems possible. Without doubt, however, the identification and description of the modes of action in biological terms is of increasing interest. Regulation by cell death, nuclear anchorage, physical barricade, attractive "forces" (Swiezynski, 1961a), and shutoff of motive force are among a diversity of possible modes under consideration. Explaining any example of regulation in molecular terms is strictly for the future, but it will be especially interesting to know if the possibilities ever reduce to any generalities for filamentous fungi.

Lethal cytoplasmic reactions could regulate nuclear migration by leaving one or more vacated cells as a physical barrier at any fusion point between incompatible strains. If the reaction is quite swift and intense, it could prevent heterokaryosis as well as migration, with no possibility even for forced heterokaryosis or for internal cross-feeding in a mosaic of homokaryons.

Although nuclear migration, as such, seems never to have been studied in any detail in *Neurospora*, the lethal cell reactions of the A/a mating system (Garnjobst and Wilson, 1956) and of the C/c and D/d heterokaryon-restricting system (Garnjobst, 1953, 1955; Garnjobst and Wilson, 1956; Wilson, Garnjobst and Tatum, 1961) of that organism probably regulate migration indirectly. The physical barriers are not altogether preventive of heterokaryosis nor of some nuclear migration, however, as the time required to kill the fused cells is too great. Within the 15—45 minutes required to kill the fusion cell, or, at least, to plug the pores fast at both ends of a fusion cell, a certain amount of protoplasm is seen to pass between the mycelia.

Since all the cells receiving this foreign protoplasm do not invariably die, it is possible that a mosaic of incompatible homokaryons could survive marginally under selective conditions by balancing the tendency to increase complementation against the tendency to destroy heterokaryotic cells. The A/a, C/c or D/d heterokaryons of $N.$ $crassa$ could be somewhat of this character.

Septal pores offer another means of regulating migration, either independently or in conjunction with lethal reactions. If the existence of septal pores is general in fungi, then means for plugging them must be also, for the plugging of pores seems to serve several important purposes. If a hypha of $N.$ $crassa$ is punctured, the nearest pores become plugged tight in a matter of $seconds$. Often only one cell dies. Plugging also cuts off protoplasmic continuity between old and vigorous portions of the mycelium. Differently designed pores could operate as shut off valves, as filter valves, or as one-way valves.

The complex septal pore of many Basidiomycetes offers intriguing possibilities. The pore has flanges termed septal pore swellings and perforated double-membranous hemispheres, called septal pore caps, over either end of the pore canal. The survey of MOORE and MCALEAR (1961) and the details of development (BRACKER and BUTLER, 1963) have provided considerable basic information about these remarkably complex pores. Electron micrographs of BRACKER and BUTLER (1964) show the pore flanges are elastic, as the pore canal doubled in diameter while accommodating a large clump of mitochondria. Structures that are probably nuclei are seen in the pore canal in other micrographs (BRACKER and BUTLER, 1964). Septal pore swellings and caps are characteristic of both homokaryotic and dikaryotic mycelia of $Schizophyllum$ (SNIDER, 1963). The suggestion of MOORE and MCALEAR (1961) that the pore structure of homo- and dikaryotic mycelia may differ markedly and characteristically has thus not been confirmed in subsequent studies. Any characteristic difference in septal structure in the two types of mycelia must be subtle. Nevertheless, these pores are probably more than fancy decorations, and regulation by them of nuclear migration as well as other sorts of transport seems highly probable.

C. Conclusion

Direct observation of the effects of migrating nuclei, such as that afforded by the mutant $puff$, or kinetic evidence which indicates that migration outdistances intrusive growth offer the most reliable means of detecting nuclear migration. In the latter instance, control experiments based upon unilateral mating can help rule out the possibility of intrusive growth. Ways of combining these methods of detection with direct observations in the light and electron microscopes open important new possibilities, as discussed by Dr. DAY and Dr. GIRBARDT in other sections of this symposium (p. 71). The most reliable values to date for the rate of nuclear migration are undoubtedly the estimates of maximal slopes from curves relating distance to time of moving migratory fronts in reticulate recipients.

Aside from new techniques, those already well developed are capable of delivering much more information of importance about migration and its regulation than is now on hand. The general properties of migration must be better understood for the formulation of testable hypotheses about the mechanism of nuclear migration and the regulation of the process.

Prevention of nuclear migration into newly formed dikaryotic cells, although not yet attributable to incompatibility genes, is probably an important type of regulation. The generality that tetrapolar Basidiomycetes have only 2 kinds of nuclei in any fruit body, a generality discussed by Dr. Burnett elsewhere in this symposium, can probably be explained now as largely the result of dikaryosis regulating against migration, so that local multikaryosis is prevented.

Regulation of the rate of migration is the first likely effect of incompatibility-gene action in Basidiomycetes to come within reach of quantification, but few reliable rate estimates exist as yet. $B=$ factors probably do not entirely block migration when other conditions are most favorable for migration, just as heterokaryosis is not entirely excluded in any combination of incompatibility factors under special conditions that force heterokaryosis. Data do not yet exclude the possibility that nuclear migration is regulated *quantitatively* by the A and B factors, with the effect of B simply more intense than that of A. If this proves correct, other quantitative effects of A and B jointly or separately are strongly implied.

Acknowledgments

The author accepts the full responsibility for the content of this paper but wishes to express appreciation to Professor John R. Raper, with whom this research was begun, to Professor Ralph Emerson for his interest in the progress of most of the research reported here, which was done during the author's tenure on the faculty at the Univ. of California, Berkeley, and especially to Dr. Peter Day, who read the manuscript and offered numerous helpful suggestions.

References

Allen, R.D., and N. Kamiya: Primitive motile systems in cell biology. New York: Academic Press 1964.

Bracker, C.E. and E.E. Butler: The ultrastructure and development of septa in hyphae of *Rhizoctonia solani*. Mycologia **55**, 35—58 (1963).

—— Function of the septal pore apparatus in *Rhizoctonia solani* during protoplasmic streaming. J. Cell. Biol. **21**, 152—157 (1964).

Brodie, H. J.: Tetrapolarity and unilateral diploidization in the bird's nest fungus, *Cyathus stercoreus*. Am. J. Botany **35**, 312—320 (1948).

Buller, A.H.R.: Researches on fungi, Vol. IV. London: Longmans and Green 1931.

Crowe, L.K.: The exchange of genes between nuclei of a dikaryon. Heredity **15**, 397—405 (1960).

— Competition between compatible nuclei in the establishment of a dikaryon in *Schizophyllum commune*. Heredity **18**, 525—533 (1963).

DODGE, B.O.: The mechanics of sexual reproduction in *Neurospora*. Mycologia **27**, 418—438 (1935).

DOWDING, E.S.: Nuclear streaming in *Gelasinospora*. Can. J. Microbiol. **4**, 295—301 (1958).

—, and A. BAKERSPIGEL: The migrating nucleus. Can. J. Microbiol. **1**, 68—78 (1954).

—, and A.H.R. BULLER: Nuclear migration in *Gelasinospora*. Mycologia **32**, 471—488 (1940).

ELLINGBOE, A.H.: Nuclear migration in dikaryotic-homokaryotic matings in *Schizophyllum commune*. Am. J. Botany **51**, 133—139 (1964).

—, and J.R. RAPER: The Buller phenomenon in *Schizophyllum commune:* nuclear selection in fully compatible dikaryotic-homokaryotic matings. Am. J. Botany **49**, 454—459 (1962).

FULTON, I.: Unilateral nuclear migration and the interaction of haploid mycelia in the fungus *Cyathus stercoreus*. Proc. Natl. Acad. Sci. (Wash.) **36**, 306—312 (1950).

— Nuclear migration and the interaction of haploid mycelia in the fungus *Cyathus stercoreus*. Dissertation, Indiana University (Bloomington) 1950.

GARNJOBST, L.J.: Genetic control of heterocaryosis in *Neurospora crassa*. Am. J. Botany **40**, 607—614 (1953).

— Further analysis of genetic control of heterokaryosis in *Neurospora crassa*. Am. J. Botany **42**, 444—448 (1955).

—, and J.F. WILSON: Heterocaryosis and protoplasmic incompatibility in *Neurospora crassa*. Proc. Natl. Acad. Sci. (Wash.) **42**, 613—618 (1956).

KIMURA, K.: Diploidisation in the Hymenomycetes. I. Preliminary experiments. Biol. J. Okayama Univ. **1**, 226—233 (1954).

MACDONALD, J.A.: The heather rhizomorph fungus, *Marasmius androsaceus* (FR.). Proc. Roy. Soc. Edinburgh **63**, 230—241 (1949).

MIDDLETON, R.B.: Evidences of common-AB heterokaryosis in *Schizophyllum commune*. Am. J. Botany **51**, 379—387 (1964).

MOORE, R.T., and J.H. MCALEAR: Fine structure of mycota. 7. Observations on septa of Ascomycetes and Basidiomycetes. Am. J. Botany **49**, 86—94 (1961).

PAPAZIAN, H.P.: Physiology of the incompatibility factors in *Schizophyllum commune*. Botan. Gaz. **112**, 143—163 (1950).

— The incompatibility factors and a related gene in *Schizophyllum commune*. Genetics **36**, 441—459 (1951).

PARAG, Y., and J.R. RAPER: Genetic recombination in a common-B cross of *Schizophyllum commune*. Nature **188**, 765—766 (1960).

PRÉVOST, G.: Étude génétique d'un Basidiomycète: *"Coprinus radiatus"* FR. ex BOLT. Thèses, Université de Paris 1962.

RAPER, J.R.: Tetrapolar sexuality. Quart. Rev. Biol. **28**, 233—259 (1953).

—, and P.G. MILES: The genetics of *Schizophyllum commune*. Genetics **43**, 530—546 (1958).

—, and J.P. SAN ANTONIO: Heterokaryotic mutagenesis in Hymenomycetes. I. Heterokaryosis in *Schizophyllum commune*. Am. J. Botany **41**, 69—86 (1954).

SANFORD, G.B., and W.P. SKOROPAD: Distribution of nuclei in hyphal cells of *Rhizoctonia solani*. Canad. J. Microbiol. **1**, 412—415 (1955).

SANSOME, E.R.: Heterokaryosis, mating-type factors, and sexual reproduction in *Neurospora*. Bull. Torrey Botan. Club **73**, 397—410 (1946).

SHATKIN, A. J., and E. L. TATUM: Electron microscopy of *Neurospora crassa* mycelia. J. Biophys. Biochem. Cytol. **6**, 423—426 (1959).

SNIDER, P. J.: Genetic evidence for nuclear migration in Basidiomycetes. Genetics **48**, 47—54 (1963).

—, and J. R. RAPER: Nuclear migration in the Basidiomycete *Schizophyllum commune*. Am. J. Botany **45**, 538—546 (1958).

SWIEZYNSKI, K. M.: Migration of nuclei in tetrapolar Basidiomycetes. Acta Soc. Botan. Polon. **30**, 529—534 (1961 a).

— Exchange of nuclei between dikaryons in *Coprinus lagopus*. Acta Soc. Botan. Polon. **30**, 535—552 (1961 b).

—, and P. R. DAY: Heterokaryon formation in *Coprinus lagopus*. Genet. Res. Camb. **1**, 114—128 (1960 a).

— — Migration of nuclei in *Coprinus lagopus*. Genet. Res. Camb. **1**, 129—139 (1960 b).

TAKEMARU, T.: Genetical studies of fungi. X. The mating system in Hymenomycetes and its genetical mechanism. Biol. J. Okayama Univ. **7**, 133—221 (1961).

WILSON, J. F., L. GARNJOBST, and E. L. TATUM: Heterocaryon incompatibility in *Neurospora crassa*—micro-injection studies. Am. J. Botany **48**, 299—305 (1961).

Discussion

Burnett: In *Schizophyllum*, when you get dikaryotization, do you get centers of dikaryotic mycelium arising at points around the periphery as opposed to uniform dikaryosis? This might have bearing on your flow hypothesis.

Snider: Yes. Patches are observed in radiate residents as you just described. The observations fit the flow hypothesis here.

Burnett: When you investigate migration in *Polystictus versicolor* you can examine the cross $A1\ B1 \times A2\ B2 \times A2\ B1$. When you examine these reciprocal crosses, you find different rates of migration. Have you seen this phenomenon?

$\rightarrow$

Plate I

a—b General plan of the nuclear migration experiments based upon recovery of donor nuclei by specific nutritional selection.
  a. A reticulate recipient mycelium fills this 15-cm petri plate. Donor implant (center); holes left after removal of test plugs appear as black spots.
  b. Growth from test plugs indicates the presence of donor nuclei. Plugs which failed to grow (7 here) beyond a minimal, allowed amount of background growth presumably lack donor nuclei.

c—f Nuclear migration appears to be restricted or blocked in radiate, homokaryotic $B=$ pairings of *S. commune* on complete medium.
  c. Fully compatible; d. common-*B*; e. common-*A*; and f. common-*AB*.

g—j Nuclear migration demonstrated in all 4 types of pairings in *S. commune*, reticulate recipients on migration complete medium. The data shown are representative of numerous results. The 8 test plugs from each of the 4 radii in the recipient mycelium are arranged here in parallel across 2 plates of selective medium, so that migration is indicated from left to right. Intervals between the vertical rows of plugs represent equal distances (5 mm) of migration.
  g. Fully compatible; h. common-*B*; i. common-*A* (The poor growth characteristic of the common-*A* heterokaryon in *S. commune* makes scoring difficult, but close inspection should show that growth occurred from all except 7 of the plugs.); j. common-*AB*.

k—n No nuclear migration was demonstrated into dikaryotic mycelia. See text.
  k. All plugs were negative from the wholly dikaryotic recipients.
  l—n Dikaryotizing migratory fronts moving toward one another met and stopped permanently at the positions indicated for each treatment.

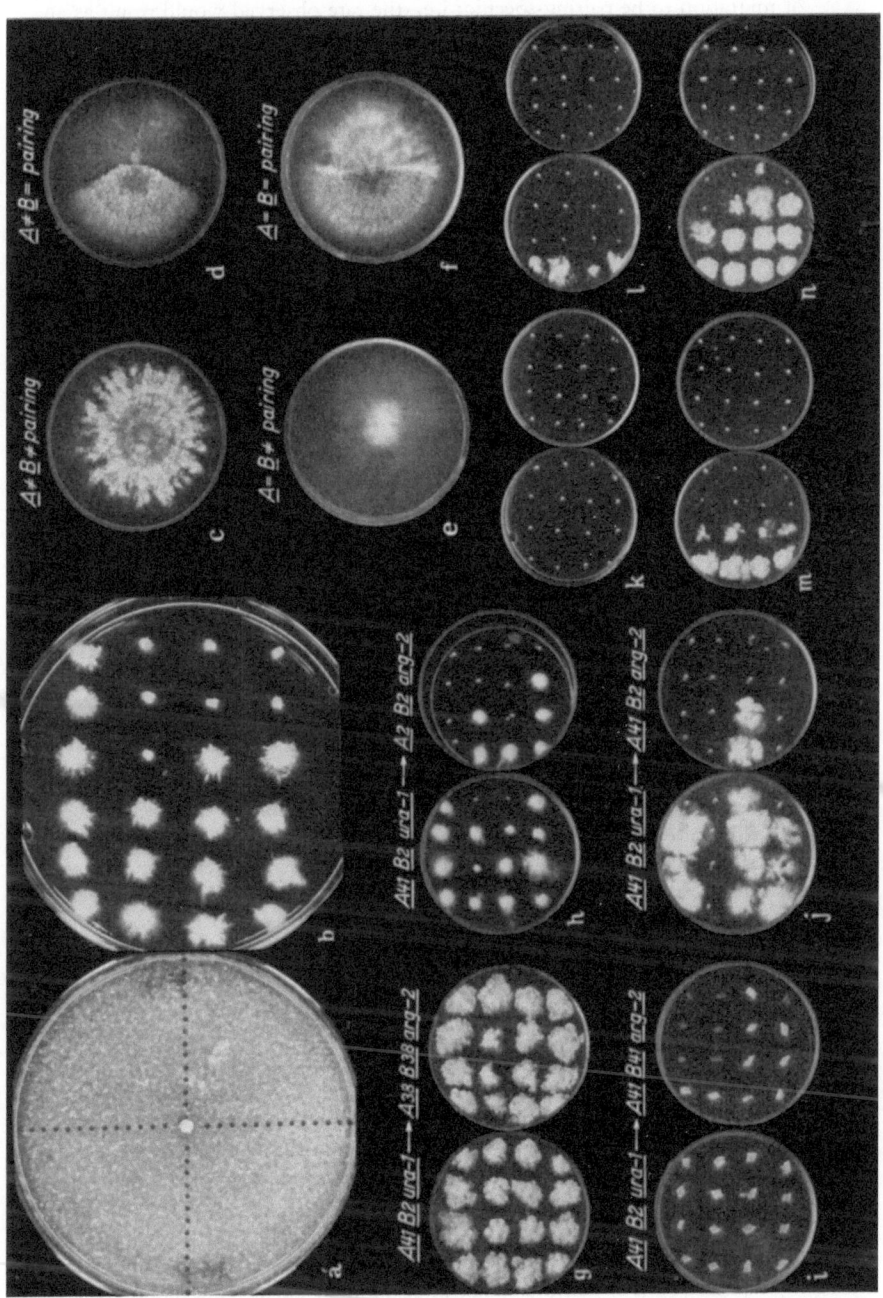

Snider: No. We haven't looked for it. The flow hypothesis would expect the rate of migration to be pairing-specific; i.e., the rate observed would result as an interaction between the two specific strains of each pairing.

Parag: I am very skeptical about nuclear migration in common-B heterokaryons. I have worked with common-B matings for years and I have never observed any significant nuclear migration. Did you not find any genetic change which might be responsible for the positive results? Could the evidence for nuclear migration in the common-*B* pairings be explained as selection of *B* mutants or of mutant modifiers of *B*?

Snider: The numerous positive plugs were spot-checked. Some were homo-karyotic for uracilless (a migrant marker), others grew on minimal medium (*B*= heterokaryons?), but none was dikaryotic (i.e., no true clamps, dika-ryotic gross morphology, or fruits). Strong selection for dikaryosis is expected in *B* = pairings, but clamps should appear also. Your skepticism is under-standable and calls for confirmation of migrant *B*-specificities recovered from a large sample of positive plugs. This has not been done. Perhaps the safest conclusion from my results is that the regular appearance of numerous posi-tive plugs from most of the cultures was reproducible in *B* = pairings prepared from *reticulate* residents.

Results of electron microscope work on Coprinus

by

P. R. DAY and R. M. GIESY

The septal pores of homokaryotic and dikaryotic hyphae of *Coprinus lagopus* as seen from ultra-thin sections are like the published descriptions from certain other Basidiomycetes. The edge of each septal pore is thickened enclosing a narrow tubular passage .09—.18 μ in diameter. On either side of the pore is a dome shaped, perforated pore cap. Both septa of the clamp connection in a dikaryon and the septa of homokaryons have pores of this type.

However, in heterokaryotic mycelia in which nuclear migration was believed to be taking place, simple pores from .4 to 1.2 μ in diameter were observed in addition to the complex pores. Rarely nuclei were observed apparently passing through these simple pores. At the same time pores were seen in various inter-mediate stages interpreted as steps in the breakdown of complex to simple pores.

The appearance of simple pores was only observed when mycelia carrying different B factors were mated. It was concluded that the association of different B factors during the establishment of a dikaryon or common A heterokaryon may bring about the breakdown of complex pores in scattered hyphae to facilitate nuclear migration.

For a full account of this work see GIESY, R. M., and P. R. DAY: The septal pores of *Coprinus lagopus* (Fr) Sensu BULLER in relation to nuclear migration. Am. J. Botany **52**, 287—293 (1965).

Discussion

Snider: Simple pores of large diameter would certainly accelerate migration ----as viewed by the flow hypothesis -----and it is probably an acceleration of rate that must be explained between $B \neq$ and $B =$ pairings. It does not seem *necessary*, however, to assume that the pore must be simple for a nucleus to pass through. BRACKER and BUTLER (1964) show a bundle of mitochondria, and separately, a nucleus straddling intact pores *and* the openings in the pore cap. The data presented for migration in $B =$ pairings are constistent with the interpretation that both fewer donor nuclei and lower rates are involved in $B =$ than in $B \neq$ pairings. Although measurement of actual rates for $B =$ pairings have yet to be made (the experiments reported merely detect migration), passage of migrants through intact pores in $B =$ pairings and through large, simple pores in $B \neq$ pairings could account for a substantial difference in rate.

Girbardt: In my technique, where it is possible to photograph the same cell, both before fixation and immediately after fixation in the electron microscope, it is sometimes observed that organelles, e.g., the mitochondria or cytoplasmic granules, which are half-way between septa in the living state are pushed by fixation pressure part of the way through an intact pore cap and pore. The pictures of BRACKER and BUTLER are probably fixation artifacts.

Snider: If fixation pressure can push a nucleus through an intact pore during fixation, why couldn't hydrostatic pressure do the same in the living mycelium, as assumed by the flow hypothesis?

Girbardt: There are four reasons why I don't think that nuclei can pass through intact cross-walls and pore caps at all. (The statements can be made only for *Polystictus versicolor.*)

1. We have observed living hyphae under optimal phase contrast for the past ten years and have never seen that nuclei, mitochondria, or little granules were migrating through the cross-wall. This would have been easily detectable, since the basal pole of the cell is optically nearly empty.

2. We observed the living fusion cell after fusion between two compatible monokaryons (LANGE, unpublished). There may be several nuclear divisions within the fusion cell, but migration of nuclei in both directions of the hyphae doesn't occur until the middle region of the cross-wall has been broken down. The newly formed, probably simple, pore may then easily be seen by phase contrast and is at least 1 μ in diameter.

3. Closed ER cisternae are always located in such a manner that they are going into the secondary hyphae. The nuclei are migrating inside these compartments and have to follow, therefore, the direction of these compartments.

4. It seems probable that the pore cap area is correlated with the synthesis of cytoplasmic membranes. It may be, therefore, that inside the pore cap there is a special lipoidal phase, and normally there is no mixing of this phase with the hydrated phase of the cytoplasm.

Snider: The discussion is now mainly about how small a pore a nucleus can pass through. Just as was stressed last year (SNIDER, 1963), we simply do not know. Why could not a nucleus squeeze through an intact pore that can stretch to 100—200 mμ in diameter?

Raper: In *Panus stypticus*, Dr. BERLINER has reported simple septa throughout. Clamps are produced in this species, but the septal pores are reported to be simple.

Snider: *Armillaria mellea*, on the other hand, has no clamps, but it does have the complex pores. Thus it seems as if we might expect all four possible combinations of the presence and absence of both complex pores and clamps.

Physiological aspects of tetrapolar incompatibility

by

STANLEY DICK

Our present ideas concerning the chemical basis of tetrapolar incompatibility system came from two basically distinct lines of inquiry, one genetic and the other physiological. We have already heard discussed in some detail the principal genetic features of the system: a) bifactorial control, b) the bipartite structure of the incompatibility factors, c) the requirement of genetic difference between factors for full compatibility and dikaryosis, d) the existence of hemi-compatible interactions when the requirements for compatibility are satisfied at only one of the two incompatibility factors, and e) the mimicking of hemicompatible phenotypes by mutations involving the incompatibility factors themselves.

The physiological study of sexuality and incompatibility has been approached from several levels. Thus, for example, the morphogenetic consequences of compatible interaction, dikaryosis, are well known. If we consider dikaryosis to be the hallmark of compatibility, and if we admit that two genetic regions control the full expression of dikaryosis, then an examination of the progression of events in hemi-compatible interactions, where the requirements of compatibility are met at only one, but not both, of the two factors, should give us some clues as to which portions of the dikaryotic process are controlled by each mating-type factor. We can, for this purpose, subdivide the process of dikaryosis into a series of sequential stages: 1) hyphal fusion, 2) nuclear migration, 3) pairwise association of the two nuclear types, 4) conjugate nuclear division, 5) clamp initiation, and 6) clamp fusion and completion.

It is fortunate that in some species, e. g., *Schizophyllum commune* (PAPAZIAN, 1950) and *Cyathus stercoreus* (FULTON, 1950), it is indeed possible to recognize in addition to the two extreme cases of a) full compatibility, as represented by the typical dikaryon, and b) no compatibility, as represented by the typical homokaryon and common-*AB* heterokaryon (MIDDLETON, 1964), two well-defined intermediate states of interaction in which partial compatibility is manifested. These intermediate states are the phenotypic expressions of the hemi-compatible heterokaryons, the common-*A* and the common-*B* heterokaryons. In the considerations that follow, we shall be referring primarily to the situation in *S. commune*, the most intensively investigated species of Basidiomycetes, but the *general* physiological pattern of development may well hold true for other tetrapolar Basidiomycetes as well (DAY, 1960; SWIEZYNSKI and DAY, 1960a; 1960b).

In the common-*A* heterokaryon, the requirement for compatibility is satisfied only at the *B* factor. Under these conditions, hyphal fusions occur as usual (hyphal fusions occur regardless of incompatibility factors), and, in addition, extensive nuclear migration is allowed, so that the nuclei of both mated mycelia become associated with one another in a common cyto-plasm. Nuclear association, how-ever, is not coordinated, and a majority of cells is still uni-nucleate, the ratio between the two nuclear types by no means approaching 1:1 as in the dikar-yon (RAPER and SAN ANTONIO, 1954; SNIDER and RAPER, 1958; RAPER and RAPER, 1964). Clamp connections are not initiated, and indeed, normal vegetative processes appear to be so dis-rupted by the "unbalanced com-patibility" that growth is deci-dedly subnormal. The hyphae are gnarled and bumpy, and some failure in wall synthesis allows the protoplasm to exude out of the mycelium into vesicles at intervals along the hyphal axis (PAPAZIAN, 1950; RAPER and SAN ANTONIO, 1954; RAPER, SAN ANTONIO and MILES, 1958; DICK, 1960). Under normal con-ditions, the sexual progression in common-*A* heterokaryons develops no further than this.

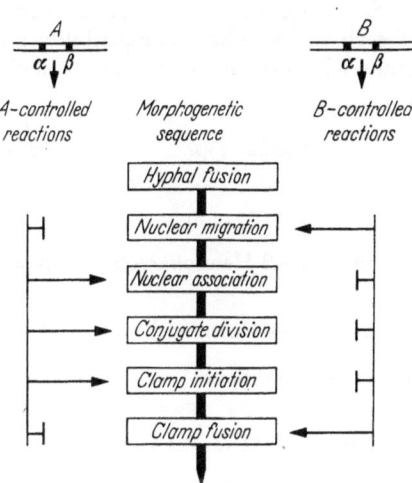

Figure 1. The control of dikaryotic processes by the *A* and *B* factors. The *A* factor, comprising two linked loci, *Aα* and *Aβ*, controls the production or coordination of enzymes responsible for nuclear association, conjugate division and clamp initiation, and the *B* factor, com-prising two linked loci, *Bα* and *Bβ*, controls in a similar fashion nuclear migration and the final stage of clamp fusion

In the common-*B* heterokaryon, the requirement for compatibility is satis-fied only at the *A* factor. Mycelia which share the same *B* factor, however, do not allow extensive nuclear migration (PARAG and RAPER, 1960). In the limited region of confluence where nuclear association does occur, a certain degree of coordinated division takes place, and clamp initials are formed (PARAG and RAPER, 1960). These pseudoclamps do not fuse with the sub-terminal cells, as do true clamps, and the clamp septation which occurs often traps one nucleus within the clamp. Although the stability of the common-*B* heterokaryon is thus not very great, hyphal tips tend to contain nuclei of both types, and the heterokaryotic state is perpetuated during the course of apical hyphal growth.

The phenotypic consequences of common factor heterokaryosis suggest a striking division of labor between the two factors in respect to dikaryosis. Accordingly, we can represent schematically the action of the two factors as controlling two series of reactions, an A series or "A pathway", and a B series or "B pathway". Although integration of the reactions under the control of both factors seems necessary for the proper coordination of certain stages of morphogenetic development, the following simplified scheme illustrates the reactions comprising the A and B pathways, respectively: the A factor is responsible for nuclear association, conjugate division and clamp initiation, and the B factor is responsible for nuclear migration and the final stage of clamp fusion (Fig. 1).

So far we know of no cases where hemi-compatibility elicits only part of each series. The incompatibility factors instead seem to act as master genes which open up the entire block of reactions under their respective

	Mycelial Type	Factor Relationships	Pathway State
a	Homokaryon	A , B	A: ⊣
b	Common-AB Heterokaryon	A = B =	B: ⊣
c	Common-A Heterokaryon	A = B ≠	A: ⊣
d	B-Mutant	A , Bm	B: ⟶
c	Common-B Heterokaryon	A ≠ B =	A: ⟶
d	A-Mutant	Am , B	B: ⊣
c	Dikaryon	A ≠ B ≠	A: ⟶
			B: ⟶

Figure 2. The effect of (a) the presence of single incompatibility factors (A, B), (b) the sharing of like incompatibility factors ($A =$, $B =$), (c) the sharing of unlike incompatibility factors ($A \neq$, $B \neq$), and (d) the presence of mutated incompatibility factors (Am, Bm) on the presumed state of operation, operative ($\rightarrow$) or inoperative ($\dashv$), of the A and B pathways in various homokaryotic and heterokaryotic mycelial types

control, and each pathway seems to exist in two states: operative and inoperative. Thus, in the normal homokaryon and in the common-AB heterokaryon, where no requirements of compatibility are met, both the A and the B pathways are inoperative. In the normal dikaryon, both pathways are operative and properly interacting. In the common-A and common-B heterokaryons, the B and A pathways, respectively, are singly operative, albeit imperfectly (Fig. 2).

In contrast to the rather detailed analysis of the morphogenetic aspects of incompatibility control, very little information has accrued concerning the biochemistry of sexuality and incompatibility. PAPAZIAN (1950) was the first to consider the interaction of compatible homokaryons in terms of chemical substances. He attempted to demonstrate the presence of hormonal substances which would induce dikaryosis in the absence of hyphal fusion between two isolates. When various homokaryotic mycelia were grown in filtrates in which compatible and incompatible homokaryons and hetero-karyons had grown, no significant changes in appearance of the mycelia were observed. Similarly, when mycelia of various types were grown on opposite sides of a cellophane membrane, no induction of heterokaryotic reactions could be seen. These experiments have been repeated, with modi-fications, by RAPER, SAN ANTONIO and MILES (1958) and recently by the author (DICK and RAPER, unpubl.), and in no case could filterable sub-stances be demonstrated that would induce the various phenotypes which are observed under the conditions of actual mating.

It appears, instead, that the intermycelial reactions that lead to hemi-incompatibility and to full compatibility are mediated by large, non-filter-able macromolecules (RAPER, 1957). These molecules ... be they proteins, polysaccharides, or nucleic acids ... are presumably the cytoplasmic bearers of the specific information resident at the A and B regions of the genome.

Polysaccharides exist in great abundance in the cells of the higher Basidiomycetes, and in view of their role in conferring specificity on bacteria and red blood cells, they were prime suspects in the search for cytoplasmic incompatibility substances. Immunochemical tests on isolated, protein-free polysaccharides, however, have shown, as yet, no consistent differences be-tween the polysaccharides of two homokaryotic strains, 699 (A41 B41) and E908 (A43 B43), which are isogenic to the extent of ten generations of backcrossing, or between the polysaccharides of the two homokaryons and the dikaryon resulting from their interaction (DICK and RAPER, un-publ.). In other words, the mycelial polysaccharides do not, in the absence of cellular proteins, reveal any involvement in the expression of mating type specificity at the molecular level.

It appears much more likely that the incompatibility factors control the production of specific *proteins*. RAPER and ESSER (1961) were the first to demonstrate specific differences between the proteins of two isogenic homo-karyons, on the one hand, and of the dikaryon formed from their interaction, on the other hand. When protein extracts of all three mycelial types were injected into rabbits, the antisera which resulted reacted in the following manner: 1) Most of the precipitin bands which formed on Ouchterlony assay plates upon the interaction of the various mycelial antigens with in-dividual antisera were present in the homokaryon and the dikaryon alike. 2) Some bands were peculiar to the dikaryon alone. Similar analyses by the author have not only confirmed these results essentially, but have also de-monstrated a band specific for at least one of the homokaryons (699). This homokaryotic band cross-reacts with a protein produced by the other homo-karyon, and thus is not very distantly related to it, but thus far there is no certain evidence of a corresponding band which is specific for the other homokaryon (E908) and which reacts reciprocally (DICK and RAPER, un-publ.).

In a further attempt to compare the proteins of the three mycelial types, acrylamide gel disc electrophoresis was used to obtain protein spectra which were independent of any antigenic activity. These analyses have also demonstrated differences not only between the homokaryons and the dikaryon, but also between the two homokaryons themselves (DICK and RAPER, unpubl.).

In the absence of any definite information as yet concerning the biochemical activity of the proteins specific to the various mycelial types, it is difficult to interpret with any great certainty the significance of these observations. Nevertheless, what can be stated is that the results are in accord with the suppositions: first, that specific proteins are associated with the dikaryotic process, these proteins presumably being those enzymes

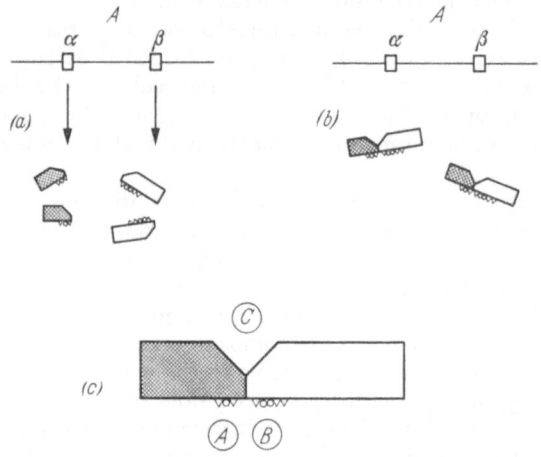

Figure 3. Diagrammatic representation of the hypothesized formation and structure of an incompatibility protein produced by an *A* factor. (a) each factor subunit (an *α* locus or a *β* locus) produces a polypeptide chain endowed with binding specificity (represented by the linear sequence of circles and triangles); (b) a non-specific dimerization of any *Aα* polypeptide with any *Aβ* polypeptide results in the formation of an intact *A* incompatibility protein; (c) the intact *A* incompatibility protein is endowed with the binding specificity of both *Aα* (Ⓐ) and *Aβ* (Ⓑ), and in addition contains a functional site (Ⓒ) which allows it to function in regulating the *A* pathway; the functional site of all *A* proteins, regardless of the mechanism of regulation, is assumed here to be identical

concerned with processes peculiar to the dikaryon, e. g., nuclear migration and clamp formation, and second, that specific proteins *might* be associated with the individual mating-types. Thus far we have been able to analyze only two homokaryotic mating-types, and it may be that other mating-types do not differ sufficiently in their incompatibility proteins for more minute differences to be detected by these methods. More effective methods of detection do exist, however, e. g., amino acid sequence analysis, and the disc electrophoretic method, particularly, will enable us to proceed with the isolation of the regions of difference and with the further analysis of these differences.

What sort of biochemical systems would allow for the genetic and physiological facts which have been presented at this symposium? The possibilities can be reduced to two: a) that the incompatibility proteins are *inactive precursor* molecules of a trigger enzyme, the *activation* of which is necessary for inducing the chain of reactions responsible for dikaryosis, and b) that the incompatibility proteins are *active inhibitors*, the *inactivation* of which is necessary for dikaryosis to proceed. In the former case, the proteins would be formally equivalent to zymogens, whereas in the latter case they could be likened to aporepressors.

Under both systems, the "incompatibility proteins", the primary products of the incompatibility factors, would have to be bipartite, i.e., dimeric, the combined product of the two subunits of each factor (Fig. 3). Moreover, under both systems, the incompatibility proteins could be endowed 1) with regions which embody the genetic specificity of the incompatibility factors themselves (Fig. 3, *binding sites* A and B), and which would determine the ability or inability of heteroallelic interaction to take place between differently constituted subunits, and 2) with an active site which enabled the proteins to act intramycelially, i.e., as repressors in the homokaryons or as activators in the heterokaryons (Fig. 3, *functional site* C).

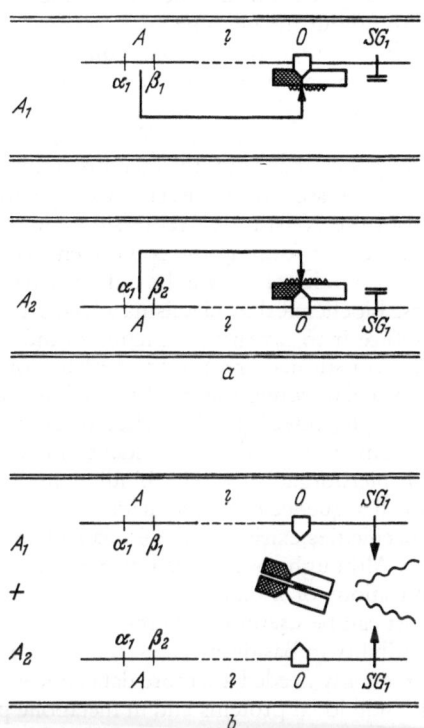

Precedents for both the inactive precursor and the active inhibitor hypotheses exist at the molecular level in work conducted on enzymatic control mechanisms, e.g., the work of TOMKINS and YIELDING (1961) on mammalian glutamic dehydrogenase, and the work of the group at the

Figure 4. Diagrammatic representation of the action of the *A* factor under a repressional inhibitional mechanism. (a) Each of two unmated homokaryons produces an *A* protein which acts as the repressor of an operator (O) which controls the activity of one or more structural genes (the first of which is indicated as SG_1) responsible for the synthesis of enzymes involved in the *A* pathway. (b) Upon mating of the two homokaryons, the *A* proteins interact by means of their binding sites (here shown to differ only at the *Aβ* region), and the release of repression allows the structural gene, SG_1, to proceed with the synthesis of its gene product. A similar mechanism would hold true for the *B* factor

Institut Pasteur on β-galactosidase of *E. coli* (see JACOB and MONOD, 1961, and MONOD and JACOB, 1961, for references). Neither hypothesis can presently be excluded for tetrapolar Basidiomycetes on either genetic or physiological grounds, but physiological economy might perhaps favor an active inhibitor mechanism. For purposes of illustration we may arbitrarily consider a mechanism of this sort (Fig. 4).

In accordance with an active inhibitor mechanism, each incompatibility factor would serve as a *regulator gene* for the series of reactions in the pathway under its control. Each of these reactions in turn would be mediated by a *structural gene*. We do not know exactly where these structural genes are located, but some evidence exists, namely the recent work of RAPER and RAPER (1964) on modifier loci, which indicates that they may be scattered throughout the genome.

In the unmated homokaryon, the product of each incompatibility factor would be a bipartite protein which acts as the repressor of an operator locus, via its *functional site* (Fig. 4a). Upon heteroallelic interaction of differently constituted *binding sites*, the sites of allelic specificity, at either of the a or β loci or at both, the incompatibility protein could be rendered inactive as a repressor, thus releasing the operator and subsequent structural genes from the constitutive repression and allowing them to proceed with the business of forming the components of the A and B pathways.

It would, of course, be quite premature to speculate further on the precise biochemical mechanisms which are involved in heteroallelic interaction. Suffice it to say that interaction of individual binding sites could inactivate the constitutive repressors perhaps by causing an internal strain in the molecular configuration. Interactions of this general sort, i.e., involving multiple, interdependent sites of activity within a single molecule, have become more and more evident in the recent analyses of allisteric proteins in microorganisms. The binding itself could be achieved by a relatively small sequence of amino acids ... 6 or 7 could account for the number of specificities exising at the β locus of the A factor.

Although space does not here permit the full exposition of all the implications of a model of this type, it is hoped that considerations of this sort can be useful in guiding further research on the physiology of incompatibility in Basidiomycetes. It should be emphasized that what is now most pressingly needed are more data concerning 1) the types of differences, both in individual proteins and in metabolic pathways, that exist between homokaryons of different mating-type and between homokaryons and dikaryons, and 2) the effects of mutations of various sorts on the incompatibility phenomena in respect to these biochemical criteria.

References

DAY, P.R.: Structure of the A mating type locus in *Coprinus lagopus*. Genetics **45**, 641 (1960).

DICK, S.: The origin of expressed mutations in *Schizophyllum commune*. Dissertation Cambridge, Mass.: Harvard University 1960.

—, and J.R. RAPER: unpublished observations.

FULTON, I.: Unilateral nuclear migration and the interaction of haploid mycelia in the fungus *Cyathus stercoreus*. Proc. Natl. Acad. Sci. (U.S.) **36**, 306—312 (1950).

JACOB, F., and J. MONOD: On the regulation of gene activity. Cold Spring Harbor Symp. Quant. Biol. **26**, 193—211 (1961).

MIDDLETON, R.B.: Evidences of common-AB heterokaryosis in *Schizophyllum commune*. Am. J. Botany **51**, 379—387 (1964).

MONOD, J., and F. JACOB: Teleonomic mechanisms in cellular metabolism, growth and differentiation. Cold Spring Harbor Symp. Quant. Biol. **26**, 389—401 (1961).

PAPAZIAN, H.P.: Physiology of the incompatibility factors in *Schizophyllum commune*. Botan. Gaz. **112**, 143—163 (1950).

PARAG, Y., and J.R. RAPER: Genetic recombination in a common-B cross of *Schizophyllum commune*. Nature **188**, 765—766 (1960).

RAPER, C.A., and J.R. RAPER: Mutations affecting heterokaryosis in *Schizophyllum commune*. Am. J. Botany **51**, 503—512 (1964).

RAPER, J.R.: Hormones and sexuality in lower plants. Symp. Soc. Exp. Biol. **11**, 143—165 (1957).

—, and K. ESSER: Antigenic differences due to the incompatibility factors in *Schizophyllum commune*. Z. Vererbungsl. **92**, 439—444 (1961).

—, and J.P. SAN ANTONIO: Heterokaryotic mutagenesis in Hymenomycetes. I. Heterokaryosis in *Schizophyllum commune*. Am. J. Botany **41**, 69—86 (1954).

— —, and P.G. MILES: The expression of mutations in common-A heterokaryons of *Schizophyllum commune*. Z. Vererbungsl. **89**, 540—558 (1958).

SNIDER, P.J., and J.R. RAPER: Nuclear migration in the Basidiomycete *Schizophyllum commune*. Am. J. Botany **45**, 538—546 (1958).

SWIEZYNSKI, K.M., and P.R. DAY: Heterokaryon formation in *Coprinus lagopus*. Genet. Res. Camb. **1**, 114—128 (1960a).

— — Migration of nuclei in *Coprinus lagopus*. Genet. Res. Camb. **1**, 129—139 (1960b).

TOMKINS, G.M., and K.L. YIELDING: Regulation of the enzymic activity of glutamic dehydrogenase mediated by changes in its structure. Cold Spring Harbor Symp. Quant. Biol. **26**, 331—341 (1961).

Discussion

Sproston: In your immunochemical and electrophoretic studies, is the pattern of the dikaryon not the same as the sum of the two homokaryons?

Dick: This has not been our experience thus far, nor would one expect it to be. After all, the dikaryon has presumably undergone several profound changes in its metabolic machinery which could not be duplicated by merely mixing two homokaryotic extracts under conditions of no protein synthesis.

Sproston: Your assumption is that proteins are involved. What is the evidence for this?

Dick: At first, the evidence was the implication of non-dialyzable substances coupled with the elimination of polysaccharides in the involvement with specificity. Right now the evidence is the positive existence of protein differences between the various strains.

Sproston: Could the incompatibility substances be steroids?

Dick: We haven't looked at steroids in *Schizophyllum*. There is lots of room for specific variation on a steroid skeleton. The question is whether there is enough.

Weresub: You mentioned that hyphal fusion occurs indiscriminately. Do you mean this for interspecific matings too?

Dick: So far as I know, I don't think that it would hold true for all interspecific matings, but it might for some.

Snider: The recent evidence for migration in common-B pairings may require further consideration of the nature of B action.

Dick: Your evidence concerning common-*B* migration was not available to us at the time these ideas were being worked out. If common-*B* migration does prove to be a general phenomenon, we may simply have to subdivide the process of migration into a *B*-controlled component and an *A*-controlled component. Even with common-*B* migration, according to what you have said, the operation of an *A*-controlled component is probably only an augmentative phenomenon under natural conditions, and relatively minor in comparison with the B-controlled component.

Snider: The rationalization of *A* as being in complete charge of clamp regulation is not compelling until the sequence is demonstrated as cause-and-effect linked. Interpreting the false clamps of $A \neq B$ = pairings as a quantitative effect of *A-B interaction* seems to be an equally tenable hypothesis.

Dick: To be sure, there are other possible hypotheses, but it seems to me that we would like to begin by investigating the simplest ones first.

Snider: Assumption of the operon concept seems the most shaky part of the model. Several *Neurospora* geneticists are among those seeking evidence for operons in eukaryotic organisms.

Dick: It is true that there are few precedents for operons in eukaryotic organisms, but the formal analogy is striking, and it ought to provide a firm basis for further testing.

Genetic investigation into the mode of action of the genes controlling self incompatibility and heterothallism in Basidiomycetes

by

YAIR PARAG

An essential problem in tetrapolar incompatibility is the means by which the specific information carried by the incompatibility genes determine compatibility or incompatibility of a given mycelium with another; this is the problem of the physical basis of the action of the incompatibility gene. A second problem is to determine which developmental stages of the sexual process are affected by the incompatibility mechanism; this is a physiological problem. The first deals with the *specific information* carried by the incompatibility genes; the other deals with the detailed *mechanisms* by which this information is translated to actual incompatibility or compatibility, regardless of the specificity of this information.

The problem of the complicated specificity governing incompatibility in the Basidiomycetes has puzzled most researchers in this field, and hypotheses to account for this complexity have been devised by many of them. It is not in the scope of this paper to review all the proposed or possible hypotheses. Short descriptions will be given of only two hypotheses, which have already been presented with almost identical details by RAPER in 1961 in a Symposium of the British Mycological Society (LEWIS, 1961). In principle, they are probably not very different from other hypotheses presented previously (LEWIS, 1954) or simultaneously and indenpendently in recent years (FINCHAM and DAY, 1963, PREVOST, 1962). The very sound hypothesis of DICK (this Symposium) will be briefly discussed later, but only in respect to the main theme of this presentation.

The objective of this paper, however, is not to describe this or another hypothesis, but rather to try to explain these theories on the basis of recent genetical findings in order to suggest genetical methods for testing these hypotheses. The discussion of these tests is based on a formal genetical presentation of the two hypotheses concerning the mode of action of the incompatibility factors. It is not claimed that the tests suggested here will definitely permit a choice between the different hypotheses — it would rather eliminate the hypotheses that are found to be inconsistent with the results of these tests.

It has been shown that the A factor and possibly the B factor of *Schizophyllum commune* are composed of two loci, α and β (RAPER, BAXTER, and MIDDLETON, 1958a; RAPER, BAXTER, and ELLINGBOE, 1960; DAY, 1960). Each locus can be separately tested in matings in which one of the loci is homoallelic. Since what we observe is the effect of the A factor as a whole, and since we want to analyse the function of the separate loci as well, consideration will sometimes be focused upon the whole factor and sometimes upon the separate loci, dependent upon their relevance. Each locus is considered, at least as a basic assumption on which we can start our discussion, to be a single gene. Some theoretical considerations supporting this assumption will be developed during this discussion.

A. The genetic formulation of the hypotheses concerning the mode of action of the incompatibility factors

Two diametrically opposite hypotheses — the *complementation* hypothesis and the *oppositional inhibition* hypothesis — will serve as the basis for this discussion.

I. The complementation hypothesis

This hypothesis assumes that "in the beginning" there was — for each locus, e.g., the $A\beta$ locus — a normal $A\beta$ allele, $A\beta0$, which was able to synthesize the substance required for dikaryotisation ("dikaryotisation substance" — DS). A series of mutations to new $A\beta$'s, $A\beta1$, $A\beta2$, etc., occurred, each leading to the inability of the $A\beta$ locus to synthesize DS. We may assume that $A\beta0$ had a specific arrangement of its coding elements, and substitution or elimination of a nucleotide pair caused a new arrangement of these elements resulting in "nonsense" information (YANOFSKY, 1960). For reasons that will become apparent later, mutations will be represented as changes in the order of numbers which denote coding elements:

$$\underline{A\beta0} \quad 1\ 2\ 3\ 4 \longrightarrow \text{Dikaryotisation substance}$$

$$\text{substitution mutation}$$

$$\underline{A\beta1} \ 1\ ⑦\ 3\ 4 \ \text{ or } \ \underline{A\beta2} \ 1\ 2\ ②\ 4 \longrightarrow \begin{array}{l}\text{no dikaryotisation}\\\text{substance}\end{array} \tag{1}$$

$$\begin{array}{l}\underline{A\beta1} \ 1\ ⑦\ 3\ 4\\[2pt]\underline{A\beta2} \ 1\ 2\ ②\ 4\end{array} \longrightarrow \text{Dikaryotisation substance}$$

When a cross involving $A\beta1 \times A\beta2$ is made, each of the mutated sites finds its matching site unmutated, and the functional order 1234 is restored, without prejudice as to level — template-template, template-product, or product-product (see Y<small>ANOFSKY</small>, 1960, and Y<small>ANOFSKY</small> and S<small>T</small>. L<small>AWRENCE</small>, 1960); DS is synthesised, and dikaryotisation takes place.

This theory has attractive features. The mechanism is very similar in appearance to that of two complementary, nutritional deficiencies, and it is very easy to rationalize the evolution of the above system from this simple genetic progression (see R<small>APER</small>, 1960a). But it has to face some serious difficulties.

1. If the A factors involved in a cross are heteroallelic for $A\beta$ but homo-allelic for Aa, non-complementation in the Aa should lead to incompatibility.

$$\begin{array}{cc}\alpha1 & \beta1\\ \text{I} & \updownarrow\\ \alpha1 & \beta2\end{array} \qquad \begin{array}{l}\vdash\!\!\dashv = \text{no complementation}\\ \longleftrightarrow = \text{complementation}\end{array} \tag{2}$$

Experimental evidence, however, clearly shows that heteroallelism at a single locus is sufficient to give a compatible reaction.

This difficulty may be overcome by an appropriate type of non-allelic interaction. The only interaction of this kind that would fit both this theory and the facts is one in which the two loci affect two alternative pathways leading to the synthesis of the dikaryotisation substance; therefore, the restoration of one pathway via complementation in heteroalleles at one locus is sufficient to complete dikaryotisation.

2. Another and more serious difficulty is the gene structure that would be required by this theory. The number of alleles in the Aa and $A\beta$ loci of *Schizophyllum commune* is estimated as 9 and 50, respectively. In a total sample of about 8000 progeny, neither intra-Aa nor intra-$A\beta$ recombination has been found (R<small>APER</small> et al., 1960). This sets the size of these loci around 10^{-4} or lower. The least frequent intergenic recombination known in the literature is 0.8×10^{-4}; the smallest gene known is 3×10^{-4} (P<small>ONTECORVO</small>, 1958). On this basis, we must assume that each locus is a single gene, or at most 2—3 genes. The exact pattern of mating interactions strongly argues against distribution of 50 alleles among two-three loci. Consequently, the 50 alleles of the $A\beta$ locus must be mutations in the same gene, complementary in all possible combinations, and presumably without any overlapping.

The claim about non-overlapping requires explanation. First, let us consider a well-investigated locus controlling a step in a biochemical pathway, e. g., *td* locus in *Neurospora*. The complementation map of *td* (YANOFSKY, 1960) shows several non-complementary alleles, and seven complementary alleles may be arranged in five *groups* (Fig. 1, after YANOFSKY, 1960).

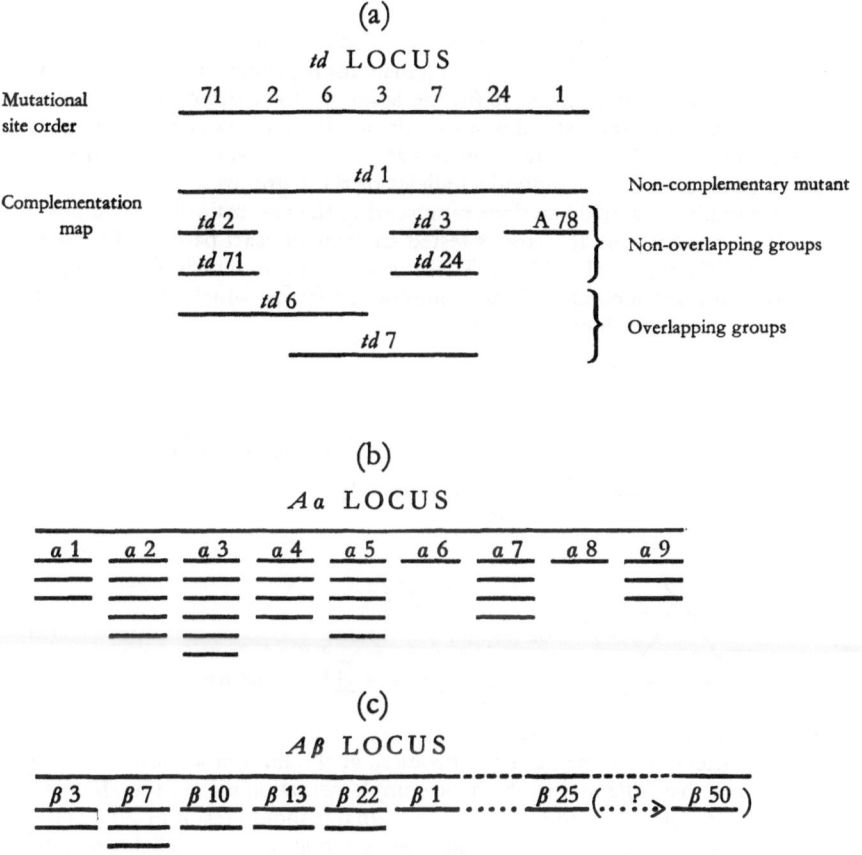

Figure 1. (a) Complementation map of *td* locus in *Neurospora crassa* (after YANOFSKY, 1960). (b) and (c) Hyptothetical "complementation maps" of *Aa* and *Aβ* loci of *Schizophyllum commune*. (Explanations in text)

The arrangement of these five *complementation groups* indicates at least four distinct *complementation regions* (WOODWARD et al., 1958). Note that the line representing *td6* overlaps those representing *td2* and *td71* as well as that representing *td7*, but the line representing *td2* does not overlap that representing *td7*. Thus, *td6* is non-complementary to both *td2* and *td7*, but *td2* si complementary to *td7*, as follows:

(3)

Such overlapping of complementation groups is usual in intragenic complementation (CATCHESIDE, 1960; FINCHAM, 1960; WOODWARD, 1960; YANOFSKY and ST. LAWRENCE, 1960; FINCHAM and DAY, 1963). The 34 alleles of the Aa and $A\beta$ loci that were independently isolated from nature are grouped into nine such "complementation groups" in the Aa locus and 25 groups in the $A\beta$ locus (in the latter, up to 50 groups are expected to be found when the natural world-wide population is exhaustively tested; RAPER et al., 1960). The question is, can we find any overlapping among these groups or among individual alleles of these groups?

A sample of ca. 6000 matings representing the vast majority of the possible pairings between the strains tested showed an exact pattern of incompatibility (RAPER et al., 1958b). There was no case in which, for example, an A factor was incompatible with another A factor, which, in turn, had a different pattern of incompatibility.

The pattern

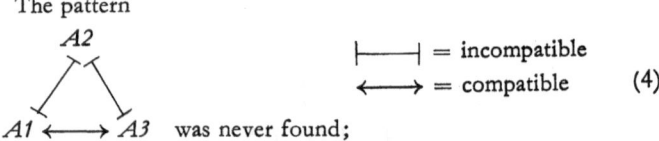

was never found;

(4)

only the patterns,

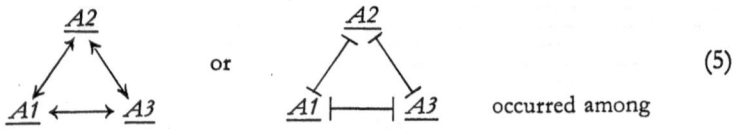

occurred among

(5)

the A factors in the sample.

It is necessary to examine the problem at the level of a single locus. If we assume that $A\beta2$ in the schematic complementation map of the $A\beta$ locus is non-complementary to both $A\beta1$ and $A\beta3$, it should result in $A\beta2$ being non-compatible with either $A\beta1$ and $A\beta3$, while these two alleles may be compatible with one another,

(6)

$A\beta1 \longleftrightarrow A\beta3$, in contradiction to scheme (5) above.

On the other hand, it can be said that the presence of different Aa's mask the existence of such an irregular pattern of incompatibility. If $Aa1$, $Aa2$ and $Aa3$ are all complementary, and $A\beta1$, $A\beta2$, $A\beta3$ have the interrelationships shown in (6), it still can be said that,

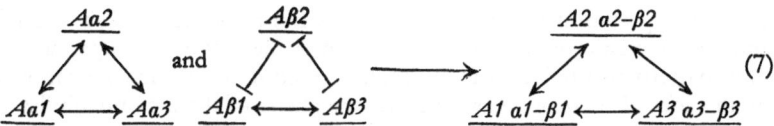

$$ (7) $$

the result would conform to scheme (5) as the observed situation. Common Aa's must, therefore, be introduced in order to analyse alleles at the $A\beta$ locus. Then,

$$ (8) $$

Such a test was actually performed (RAPER et al., 1958a, 1960). In this test, recombinants between $A41\ a1—\beta1$ and $Ax\ ax—\beta y$ were always $A\ a1—\beta y$ or $A\ ax—\beta1$. Each group of recombinants had, therefore, either $Aa1$ or $A\beta1$ in common. These recombinants were mated in many combinations. Although for methodological reasons they were not tested in all possible combinations, the results so far are suggestive enough that scheme (8) is invalid. A further test of the other possible combinations should give the decisive answer. It can be added that the number of alleles in the loci of the B factor is lower, and there should be a fair probability of finding B factors with overlapping incompatibility if they really exist, but no overlapping was found in the 6000 tests. Although the specific experiment has not been performed to give the decisive answer to whether there is such overlapping incompatibility, the mating interactions in these 6000 matings involving both factors of $S.$ *commune* (RAPER et al., 1958b) as well as those from many other tetrapolar Basidiomycetes (DAY, 1963b; FRIES, 1940, 1943; FRIES and JONASSON, 1941; LANGE, 1952; PRÉVOST, 1962) provide strong evidence for non-overlapping. Until there is experimental evidence to the contrary, this model can be considered as valid. This non-overlapping leads us to postulate the "complementation maps" of the Aa and $A\beta$ alleles of the A factor of $S.$ *commune* as given in Fig. 1.

This non-overlapping complementation map makes the complementation hypothesis rather difficult to accept on a genetic basis. There are, however, some extenuations that would make the hypothesis less unattractive. 1) It is possible that natural selection worked against the overlapping mutations, as their outbreeding efficiency would be less than that of the non-overlapping ones. For example, in Fig. 1a, if we eliminate the two overlapping groups, a non-overlapping type of complementation map would result. Such elimination of overlapping groups in the incompatibility factors might well be accomplished by nature. However, the knowledge of the *td* complementation map has been extended well beyond the state cited here (CATCHESIDE, pers. comm.), and it now appears to be so involved that such elimination should lead nowhere. 2) BERNSTEIN and MILLER (1961) found that the 378 complementary alleles fell into only five complementation groups, which they considered as the arrangement typical to this locus.

Their interpretation was that "each locus has associated with it a unique and characteristic pattern of discrete complementation" groups, and that "this pattern also reflects the molecular structure of the enzyme controlled by this locus." It may be argued that the alleles of the incompatibility loci have their characteristic pattern, a non-overlapping type, selected by nature, because it is most efficient. 3) In the house mouse, every *tailless* allele appeared capable of a certain degree of complementation one with the other (DUNN, 1956). The last case is particularly interesting, since it involves a series of alleles isolated from nature, which exist in the diploid organism only in the heterozygous condition in a "balanced lethal" system and therefore serve as a series of incompatibility alleles.

Nevertheless, the situation of *fifty* non-overlapping complementation groups is still an extreme peculiarity which makes the complementation theory difficult to accept.

II. The oppositional inhibition hypothesis

This hypothesis has been developed by LEWIS (1954) in relation to incompatibility in higher plants and has been considered in relation to incompatibility in the higher fungi by RAPER (cf. LEWIS, 1961). The rationale behind this model is that it tries to avoid difficulties encountered by several other models: there is no need for interallelic complementation; each allele is responsible for the synthesis of a single specific molecule; the product of each allele reacts only with its own product and avoids products of all other alleles in the series. It is assumed that during mating, dikaryotisation will proceed unless a positive action of incompatibility blocks it. If the mating involves, for example, *A1* and *A1*, the gene product of *A1* from one cell will interact with the gene product of *A1* from the other cell, and this positive interaction will directly prevent dikaryotisation. If the mating involves *A1* and *A2* (or *A3*, etc.), the gene products would not "recognize" each other, and the lack of any blocking interaction would permit dikaryotisation. I shall not discuss the difficulties concerning this hypothesis; I shall only say that at the moment it is very difficult to postulate biochemical entities or conditions that will enable such a mechanism to work. On the other hand, it seems to fit beautifully the available genetic information, and it offers a very convenient genetic formulation that can easily take into account many established facts.

The genic basis for the difference between alleles is specifically different codes, i.e., different orders of the coding units of the genetic material yield information for the construction of specifically different products, but all are equally active ("missense" information).

$$
\begin{array}{lll}
\underline{A\beta1} & 1\ 2\ 5\ 7 & \text{product 1} \\
\underline{A\beta2} & 2\ 6\ 4\ 5 & \text{product 2, etc.}
\end{array}
\tag{9}
$$

$A\alpha$ and $A\beta$ produce a combined product, which is responsible for the specific activity of the A factor as a whole.

Factor $A41\ a1\text{–}\beta1$ would produce protein $\boxed{1}\,\boxed{1}$

Factor $A51\ a2\text{–}\beta2$ would produce protein $\boxed{2}\,\boxed{2}$ (10)

Factor $Aa\ \ a1\text{–}\beta2$ would produce protein $\boxed{1}\,\boxed{2}$

All combined products are specifically different and carry different information. Such a mechanism could explain the fact that homoallelism for one locus only gives full compatibility, since non-identical proteins are involved.

The theory requires neither 50 loci in an $A\beta$ complex-locus nor 50 non-complementary mutations in the $A\beta$ locus, but 50 different arrangements, which is easier to accept.

The postulation of many alternative alleles of a single locus, all equally active but specifically distinct in structure, can be compared with the situation in the *tyrosinase* locus in *Neurospora*. HOROWITZ et al. (1961) tested the enzyme tyrosinase of 12 strains independently isolated from nature, and found four different enzymes. The four enzymes were equally active, but they were distinguishable by their electrophoretic mobility and thermostability. Here we find that the natural population carries several alleles of T, all of which give the information for *active* but *different* tyrosinases ("missense"). It is assumed that more than a single alteration in the sequence of coding units of the gene may be responsible for the differences among the four alleles. This is also applicable to the incompatibility factors according to the oppositional-inhibition hypothesis.

B. Genetic tests

On the basis of the formulations described above, genetic tests can be performed which might decide between these two formulations and, therefore, discriminate against some of the alternative hypotheses on which these formulations are based. These tests would be based upon 1) intra-genic recombination and 2) mutations.

I. Intra-genic recombination

According to the complementation hypothesis, crossing over between two alleles should result — as in crossing over between two biochemical deficient alleles — in only two products; one would be the "wild type" allele, the other would be a double mutant. To be consistent with this hypothesis, the segregant carrying the wild type allele must be self-fertile (as far as this locus is concerned), while the segregant carrying the doubly mutated gene must be doubly incompatible, i.e., incompatible with both parents, as follows.

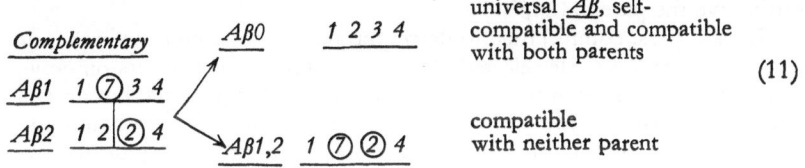

universal $A\beta$, self-compatible and compatible with both parents

(11)

compatible with neither parent

The situation according to the alternative hypothesis is different. Two *new* patterns are expected to be formed in each crossing-over event, and either each would carry "missense" information leading to a *new* active allele, i.e., to a new $A\beta$, or one or both new patterns would carry "nonsense" information leading to no information concerning incompatibility and, therefore, to universal, self-fertile $A\beta$'s. If the two alleles are different in more than two coding units, crossing over in different regions could lead to four or more different products:

Oppositional

$$
\begin{array}{lcccc}
A\beta1 & 1\ 2\ 5\ 7 & & & \\
\end{array}
\left\{
\begin{array}{llll}
A\beta3 & 1\ 6\ 4\ 5 \\
A\beta4 & 2\ 2\ 5\ 7 \\
\end{array}
\right.
$$

$$
\begin{array}{lcccc}
A\beta2 & 2\ 6\ 4\ 5 \\
\end{array}
\left\{
\begin{array}{llll}
A\beta5 & 1\ 2\ 4\ 5 \\
A\beta6 & 2\ 6\ 5\ 7,\ \text{etc.} \\
\end{array}
\right.
\tag{12}
$$

All are new $A\beta$'s. Some new patterns might be nonsense, therefore, behave like the $A\beta0$ above, i.e., universal $A\beta$'s, compatible with all $A\beta$'s and self-fertile.

The two hypotheses will offer different expectations also when the cross-over recombinants are back-crossed:

Complementary

$$
\begin{array}{llll}
A\beta0 & 1\ 2\ 3\ 4 & \longrightarrow & A\beta2 & 1\ 2\ 2\ 4 \\
A\beta1,2 & 1\ 7\ 2\ 4 & & A\beta1 & 1\ 7\ 3\ 4
\end{array}
\quad \text{parental types} \tag{13}
$$

Oppositional

$$
\begin{array}{llll}
A\beta3 & 1\ 6\ 4\ 5 & \longrightarrow & A\beta7 & 1\ 6\ 5\ 7 \\
A\beta4 & 2\ 2\ 5\ 7 & & A\beta8 & 2\ 2\ 4\ 5
\end{array}
\quad \text{new } A\beta\text{'s} \tag{14}
$$

etc.

By these tests, therefore, the complementary hypothesis will be supported if 1) only two recombinant classes are obtained, 2) one of the classes is incompatible with both parents, the other is universally compatible and self-fertile, and 3) crosses between the recombinants give again only the two parental classes. The oppositional hypothesis will be supported if 1) more than two recombinant classes are obtained, 2) all are compatible with both parental $A\beta$'s, and 3) crosses between the recombinants will not necessarily give the parental $A\beta$'s.

RAPER (pers. comm.) tried to detect intra-genic crossing over in the Aa locus of *S. commune*. He selected 5500 recombinants for markers on either side of the locus from a total sample of ca. 120,000 segregants, but no evidence for intra-locus recombination was found. Such a result is disappointing, but no clear conclusion can be drawn about the feasibility of the test.

It is possible that the altered sites of the two particular alleles used are extremely close, and crosses between other pairs of Aa alleles must be tried. It is also relevant that the intragenic recombination observed so far in loci controlling biochemical requirements is extremely low in $S.$ *commune:* in crosses involving independent mutations, *arg-1* $\times$ *arg-1* gave prototrophs in a frequency of ca. 10^{-5} (PARAG, 1960), and *ura-1* $\times$ *ura-1* gave prototrophs at a significantly lower frequency (ELLINGBOE, pers. comm.). The feasibility of crossing B-factor mutants suggests a new selective technique for the detection of intragenic crossing over (PARAG, 1962).

II. Mutations

The techniques for the selection of mutations affecting the incompatibility factors involve common-A or common-B heterokaryons, in which such a mutation should lead to dikaryosis:

$$A1 \longrightarrow Am$$
$$(A1\ B1 : A1\ B2) \longrightarrow (A1\ B2 + Am\ B2),\ \text{a dikaryon.}$$

Alternatively, it may involve homokaryons carrying mutations of the A or B factors (in which the mutations lead to self-compatibility of the affected factor):

$$A1 \longrightarrow Am$$
$$A1\ Bm \longrightarrow (A1\ Bm + Am\ Bm),\ \text{a dikaryon.}$$

Of course, by this technique, only if the new A is compatible with the original A or if it leads to self-fertility, could it be detected.

The complementary theory

A gene with a mutation in an additional site will be non-complementary to its original allele and to other alleles at the same locus. It will therefore be doubly incompatible, and, since still incompatible with the parental factors, it would be undetectible by the available methods of selection. A back mutation will restore the original, hypothetical "wild-type" allele; it would therefore not be only compatible with both parental alleles but universally compatible and self-fertile. No new Aa or $A\beta$ can be detected.

$$
\begin{array}{c}
\text{additional} \longrightarrow A\beta1,2 \quad 1\ ⑦\ ②\ 4 \\
\text{mutation} \qquad\qquad \text{incompatible with} \\
\qquad\qquad\qquad \text{the parental} \\
\qquad\qquad\qquad \text{therefore undetectible} \\
A\beta1 \quad 1\ ⑦\ 3\ 4 \\
\text{back mutation} \longrightarrow A\beta0 \quad 1\ 2\ 3\ 4 \\
\qquad\qquad\qquad \text{universally compatible} \\
\qquad\qquad\qquad \text{and self-fertile}
\end{array}
\tag{15}
$$

The oppositional inhibition theory

The mutation would be either of the "missense" type leading to a new allele or of the "nonsense" type bearing no information about incompatibility; it would therefore be universally compatible and self-fertile.

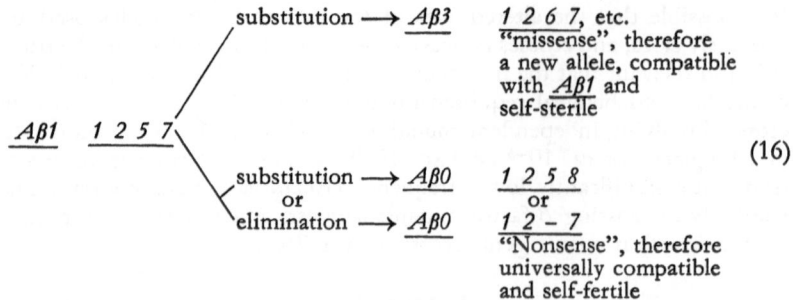

$$1\ 2\ 6\ 7, \text{etc.}$$
"missense", therefore
a new allele, compatible
with $A\beta1$ and
self-sterile

(16)

$$1\ 2\ 5\ 8$$
or
$$1\ 2\ -\ 7$$
"Nonsense", therefore
universally compatible
and self-fertile

According to the first hypothesis, only mutations for universal $A\beta$ should be detected; by the second hypothesis, new $A\beta$ specificities should be detected as well as universal $A\beta$'s. Therefore the only information which might tell which hypothesis is correct will be the isolation of *new* $A\beta$'s, and this would support the second hypothesis. The few mutations obtained so far in the B factor (PARAG, 1962; PARAG and RAPER, 1960) and the A factor (RAPER et al., 1965) of *Schizophyllum*, and the mutations in the A factor of *Coprinus lagopus* (DAY, 1963a) have all been found to be of the universally compatible class and thus could not be useful in distinguishing between the two models. Further studies on mutations in the incompatibility loci are clearly needed.

C. On the universality of the incompatibility factors: the relation between tetrapolarity, bipolarity and homothallism

The foregoing section is based on the supported assumption that the A and B factors have the same genetic role in all the tetrapolar Basidiomycetes (see also: DAY, this Symposium; DICK, this Symposium). It is also of interest that the genetic control of heterothallism by genes with multiple series of alleles is confined in fungi only to the higher Basidiomycetes — both bipolar and tetrapolar systems (RAPER, 1953, 1959a; WHITEHOUSE, 1949). It is therefore inescapable to look upon this system not only as a unique one but also as a unified system. What genetic factors determine the incompatibility in the bipolar system? We may consider three alternative propositions.

1) It is very likely that the bipolar species of the Basidiomycetes originated from tetrapolar ones (RAPER, 1953, WHITEHOUSE, 1949). If this is correct, it is not improbable that it happened by a "loss" of the function of one of the incompatibility factors (a phenomenon known from laboratory experiments: DAY, 1963a; PARAG, 1962; RAPER et al., 1965), and a compatible cross in bipolar species can be written $A1\ B0 \times A2\ B0$ or $A0\ B1 \times A0\ B2$. Since the function of A or B is universal, it might be expected that in incompatible matings in bipolar forms either common-A or common-B reactions should be recognized. As has been pointed out by RAPER (pers. comm.), however, the morphology of a common-A (or common-B) heterokaryon should thus appear also in the homokaryon, and this is very un-

likely. Recent studies by FLEXER (1963) with the bipolar fungi *Polyporus betulinus* and *P. palustris* have not revealed any discernible mating reaction similiar to the common-*A* and common-*B* interactions of tetrapolar forms. On the other hand, the loss of function by mutation of either the *A* or the *B* factor in tetrapolar species resulted in homokaryotic morphology identical with common-*B* and common-*A* heterokaryons, respectively. All of this suggests that other genes might also be operative as the background for bipolarity. A possible — at least hypothetically — way to test the hypothesis

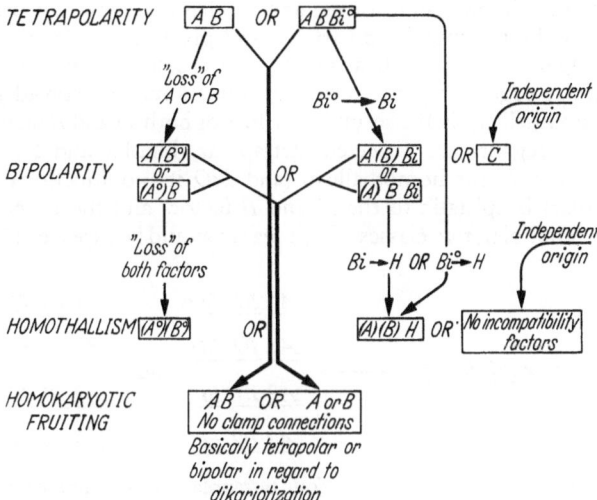

Figure 2. Schematic representation of the possible interrelationships between tetrapolarity, bipolarity and homothallism in the higher Basidiomycetes (Hymenomycetes and Gastromycetes). *A, B, C*-Incompatibility factors. $A°,B°$-Lost or inactive non-functional incompatibility factors. $Bi°,Bi,H$-Genes (allelic) controlling tetrapolarity, bipolarity and homothallism, respectively. $(A),(B)$-Normal, theoretically functional incompatibility factors, inactive due to the presence of the epistatic gene(s) *Bi* or *H*

that the bipolar incompatibility factor is identical with the *A* or the *B* of the tetrapolar system, is to intermate closely related species, one tetrapolar and the other bipolar. Such a cross could be written *A1 B1* × *A2 B0* (or *A0 B2*) and should result in the transfer of a single factor, the *A* (or *B*), into the tetrapolar genome, and part of the progeny should be *A2 B1* (or *A1 B2*).

2) It is possible that a bipolar organism carries both *A* and *B* factors, but they are masked by a third, epistatic factor, which controlls bipolarity and converts a basically tetrapolar organism into a bipolar one. In a cross between bipolar and tetrapolar, *A1 B1 Bi* × *A2 B2 Bi0* (*Bi*, a gene controlling bipolarity and non-functional, i.e., *Bi0*, in the tetrapolar), one class of segregants should be *A1 B1 Bi0*, i.e., tetrapolar segregants carrying the *A* and *B* factors previously concealed in the bipolar mate.

3) The third possibility is that bipolarity and tetrapolarity have different origins and the incompatibility factor *C* of bipolar organisms has no relation

to and possibly no similar function with the A and B factors of tetrapolar forms.

Adequate organisms with which to attempt such tests might be members of the genus *Coprinus*, with its many species and high diversity of incompatibility mechanisms, from homothallism through bipolarity to tetrapolarity, even in the same species (LANGE, 1952).

And what about the homothallic species? This group may be divided into two, exclusive of secondary homothallic forms (WHITEHOUSE, 1949; RAPER, 1960a). The larger group (QUINTANILHA and PINTO-LOPES, 1950) comprises truly homothallic forms, and in these a mycelium originating from a single basidiospore is dikaryotic and gives rise to fruiting bodies. Assuming again the hypothetical situation of crosses between homothallic and tetrapolar organisms, two main possibilities may be considered here.

a) If homothallism is the result of the loss of both A and B factors, three classes of segregants are expected: tetrapolar, bipolar and homothallic.

b) If a gene H for homothallism (and BiO its nonfunctional allele in the tetrapolar) is epistatic to the A and B factors and the homothallic is cryptically tetrapolar, two classes of segregants would be expected as follows:

$$\underline{A1\ B1\ Bi^{\circ}} \times \underline{A2\ B2\ H} \longrightarrow
\begin{array}{c}
\underline{A1\ B1\ Bi^{\circ}} \\[2pt]
\underline{A1\ B2\ Bi^{\circ}} \\[2pt]
\underline{A2\ B1\ Bi^{\circ}} \\[2pt]
\underline{A2\ B2\ Bi^{\circ}}
\end{array}
\quad \text{and} \quad
\begin{array}{c}
\underline{A1\ B1\ H} \\[2pt]
\underline{A1\ B2\ H} \\[2pt]
\underline{A2\ B1\ H} \\[2pt]
\underline{A2\ B2\ H}
\end{array}
\qquad (17)$$

four tetrapolar   four homothallic
types   types

phenotypically
indistinguishable

Although no interspecific matings have so far been successful (LANGE, 1952; RAPER, 1960b), LANGE (1952) reported that in studies with two species of *Coprinus*, *C. subpurpureus*, and *C. plagioporus*, single fruit bodies gave segregants of two classes: a tetrapolar class and a homothallic class in approximately equal frequency. This situation, termed by him *amphithallism*, I should like to interpret as a segregation for alleles controlling homothallism *vs.* tetrapolarity in addition to the A and B factor as represented in scheme (17). If this interpretation is correct, then a) fertile crosses between homothallic and tetrapolar segregants should give again the same segregation, and b) if the tetrapolar segregants from a single fruit body were used as mating-type testers, the tetrapolar strains from the same fruit body should show a tetrapolar pattern of mating interactions. A further test of the segregation in *C. plagiosporus* revealed, however, that the self-fertile ("homothallic") segregants were already dikaryotic and heteroallelic for the two incompatibility factors — i.e., another case of secondary homothallism (LAMOURE, 1957). Such a test has not yet been carried out with the other "amphithallic species", *C. subpurpureus*, described by LANGE (1952).

While so far there is no successful experiment in this line in the Basidio-mycetes, a phenomenon of a gene for homothallism masking existing mating types is known in yeasts. WINGE and ROBERTS (1949) found by crossing the homothallic *S. chevalieri* with *S. cerevisiae* that the first is basically bipolar with a gene for self-dipoidisation, which converts a genotypically bipolar form to a phenotypically homothallic one. TAKAHASHI (1958) described several genes for self-diploidisation in *S. cerevisiae*, and these appear via mutations in bipolar strains. Crosses between the bipolar (species or strains) and the homothallic (species or strains) gave two classes of segregants, homothallic and bipolar, similar to the expectation according to scheme (17).

In the second group (QUINTANILHA and PINTO-LOPES, 1950), no clear dikaryotic mycelium with clamp connections is found, and there is still the question whether, or to what extent, they may be considered as homothallic. The fruit bodies of this second group (*Psalliota campestris* — COLSON, 1935; KLUSHNIKOVA, 1939; SARAZIN, 1955; *Mycena spp.* — SMITH, 1934; KÜHNER, 1938; *Schizophyllum umbrinum* — RAPER, 1959b; and others — SASS, 1929; QUINTANILHA and PINTO-LOPES, 1950) can be com-pared to homokaryotic fruiting bodies of *S. commune*, in some strains of which species they are frequent (RAPER et al., 1958; PARAG, 1960) and of *Collybia velutipes* (ZATTLER, 1924). Usually the homokaryotic fruiting bodies of *S. commune* are small and produce few spores of very low viability, but at least in one case, large, healthy-looking homokaryotic fruiting bodies shed abundant spores having normal viability (PARAG, 1960). Rare fruiting bodies in common-*B* heterokaryons of *S. commune* have been demonstrated to complete the sexual process and thus to overcome the incompatibility barrier (PARAG, 1960; PARAG and RAPER, 1961). Recently MIDDLETON (1964) has demonstrated genetic recombination in fruiting bodies emerging from common *AB*, e. g., *A1 B1* × *A1 B1*, heterokaryons. Such fruiting bodies can be equated with homokaryotic fruiting bodies as far as the incompatibility factors are concerned. It can therefore be claimed that genetic recombination does take place in homokaryotic fruit bodies. The species with homothallic fruiting bodies mentioned above may be different only quantitatively from *S. commune*. The homokaryotic fruiting body as the site of genetic recombi-nation becomes the rule rather than the exception. Imagine only the more common appearance of such fruiting bodies in homokaryotic mycelia, and we have the situation which exists in *P. campestris*, *S. umbrinum*, etc. The evidence for meiosis in homokaryotic fruiting bodies may suggest that these homothallic species have a life cycle complete with the essential sexual events, karyogamy and meiosis, with the elimination of the dikaryotic stage as well as the self-sterility that existed in their ancestral forms.

A general model can be suggested for the relationship between the sexual types of the higher Basidiomycetes (Fig. 2). The course of events, i. e., the direction of the arrows, has been set according to the hypothesis that the tetrapolar system is the basic one from which the others evolved — the hypothesis considered by the author as the most likely one. It is, how-ever, irrelevant here. Acceptance of another evolutionary trend would only shift the directions of the arrows. The main point in this scheme is its summary of some of the possible interrelationships between the genetic factors controlling incompatibility or self-sterility in the different sexual systems of the higher Basidiomycetes.

D. Summary and conclusions

This paper deals with some hypotheses concerning the mode of action of the incompatibility factors of *S. commune* and of tetrapolar organisms in general. The main concern in this was to examine the genetic basis of the specificity of the action of the *A* and *B* factors and their component loci. Genetic formulation was given to two of these hypotheses: the complementation hypothesis and the oppositional-inhibition hypothesis. On the basis of these formulations, genetic tests are suggested that might distinguish between the different formulations and, therefore, between the corresponding hypotheses. These genetic tests are based on intragenic crossing over and on mutations in the genes of the incompatibility factors. Each of these hypotheses suggests either a unique system of interallelic complementation in the complementary hypothesis or unique types of intragenic recombination giving more than two possible reciprocal classes of recombinants in the oppositional hypothesis. Either may, therefore, serve as an interesting way for investigating the fine structure of the genetic material, and so the incompatibility system of the Basidiomycetes may serve genetic research well beyond the investigation of sexuality only.

Not all the possible tests now at hand are discussed (e.g., mutations that modify the control exerted by *A* and *B* factors, as described by RAPER and RAPER, 1964, and by DAY, 1963a). Also, the author does not claim to analyse all the possible ways to interpret the results as may be hoped to emerge from these genetic tests, but it is believed that the line of thought represented here may encourage research in this direction as well as suggest the type of research which needs to be performed and the significance of this research.

Analysis of the typical mating reactions of semi-compatible matings suggests the targets of the actions of the incompatibility factors, i.e., the stages of the sexual process affected. This analysis leads also to a suggested relationship between tetrapolar incompatibility, bipolar incompatibility, and homothallism. Genetic tests are also suggested for testing this relationship.

Addendum: repressional inhibition model

In this symposium, Dr. DICK presents in some detail a *repressional inhibition* model for the action of incompatibility factors. This model has both similarities and dissimilarities with the oppositional inhibition theory. Like the latter, DICK's model suggests that all "allelic proteins" controlled by the incompatibility alleles are functionally the same but specifically different from each other. In contrast to the oppositional inhibition theory, however, it requires (a) a positive interaction of each protein with all "allelic proteins" except itself; (b) this interaction releases the existing inhibition of dikaryosis (inhibition controlled by another, non-specific site of the same protein). Since the models are different and I have suggested that genetic tests should distinguish between different models, it is germane to ask whether this model should lead to different predictions in the genetic tests outlined above.

1) *Crossing over:* If crossing over occurs in the region controlling $A\beta$-$A\beta$ interaction, scheme (12) should apply. We should expect to find many recombinant classes. Those that lead to missense will give new $A\beta$ specificities. Those that lead to nonsense, however, should prevent the $A\beta$-$A\beta$ interaction altogether. Since this should lead to no release of the inhibition, such recombinants should be self-sterile and incompatible with all $A\beta$'s.

2) *Mutation:* Here, a theoretical distinction should be made between mutations (a) in the region controlling the $A\beta$-$A\beta$ interaction, i.e., allelic specificity, and (b) in the region controlling repression. In (a), the region conferring specificity, scheme (16) should apply, and here again, as in crossing over, missense mutations should lead to new $A\beta$'s, while nonsense mutations should be self-sterile and unable to mate with any strain. In (b), the region controlling repression, the only recognizable mutations would probably be nonsense, with the consequent de-repression of function leading to self-fertility.

All of the primary mutations in the incompatibility loci that have been recovered to date (PARAG, 1962; DAY, 1963; RAPER et al., 1965), however, appear to encompass both specificity and repression and presumably affect both hypothetical regions. Furthermore, all of these mutations are interactive, i.e., compatible, with their progenitor alleles, a fact that might reflect only the selective system used for their recovery. RAPER (pers. comm.) has suggested the possibility of "dominant" and "recessive" mutations and the use of mutant B strains as a likely selective system for "recessive" A-factor mutations. Phenotypically, the common-A heterokaryon and the mutant B homokaryon are indistinguishable, but the fate of "recessive" A-factor mutations would be quite different in the two systems:

$$A1 \longrightarrow A1m^r$$
$$(A1\ B1 : A1\ B2) \longrightarrow (A1\ B1 + A1m^r\ B2)$$

would remain a common-A heterokaryon because of the dominance of $A1$ over $A1m$; while

$$A1 \longrightarrow A1m^r$$
$$A1\ Bm \longrightarrow (A1m^r\ Bm + A1m^r\ Bm)$$

would yield a dikaryon, since both factors would then be self-fertile.

The recovery of such "recessive" mutations in the incompatibility loci would support the repressional inhibition model as described by DICK, and the current search for these mutations offers a promising possibility.

(Editor's Note: Subsequent to the Symposium and the submission of Dr. PARAG's paper, secondary mutations at the $B\beta$ locus of *Schizophyllum commune* have been described by RAPER et al. (1965). These mutations, derived from a $B\beta$ mutation provided by PARAG, bear out his suggestion of separable functions of specificity and repression. Each of four such mutations recovered restores normal homokaryotic morphology and self-incompatibility (repressive function) while rendering the allele compatible with its original wild progenitor allele (specificity function). The alteration of specificity could have occurred either in the primary mutation, in which it would have been masked by the universal compatibility of the mutant allele, or in the secondary mutation. That the four secondary mutations derived from a single primary mutation are incompatible with each other strongly suggests that the new specificity was determined in the primary mutation. J.R.R.)

Acknowledgments

My sincere thanks are due to Prof. J.R. RAPER for his invaluable advise during the writing of this article and for his critical reading of the manuscript. I am indebted to Professors D.G. CATCHESIDE, S. DICK, S. EMERSON, J.R.S. FINCHAM, N.H. HOROWITZ and P.J. SNIDER for the interesting discussions and suggestions, and for the permission to use experimental material unpublished at the time when the present paper was prepared.

References

BERNSTEIN, H., and A. MILLER: Complementation studies with isoleucine-valine mutants of *Neurospora crassa*. Genetics **46**, 1039—1052 (1961).

CATCHESIDE, D.G.: Complementation among histidine mutants of *Neurospora crassa*. Proc. Roy. Soc. (London), Ser. B. **153**, 179—194 (1960).

COLSON, B.: The cytology of the mushroom *Psalliota campestris*. Ann. Botany **49**, 1—18 (1935).

DAY, P.R.: The structure of the A mating type factor in *Coprinus lagopus*. Genetics **45**, 641—650 (1960).

— Mutations of the A mating type factor in *Coprinus lagopus*. Genet. Res. Camb. **4**, 55—64 (1963a).

— The structure of the A mating type factor in *Coprinus lagopus:* wild alleles. Genet. Res. Camb. **4**, 323—325 (1963b).

— The genetics of tetrapolar incompatibility. This Symposium, pp. 31—36 (1965).

DICK, S.: Physiological aspects of tetrapolar incompatibility. This Symposium, pp. 72—80 (1965).

DUNN, L.C.: Analysis of a complex gene in the house mouse. Cold Spring Harbor Symp. Quant. Biol. **21**, 187—195 (1956).

FINCHAM, J.R.S.: Genetically controlled differences in enzyme activity. Advan. Enzymol. **22**, 1—43 (1960).

—, and P.R. DAY: Fungal Genetics. Blackwell, Oxford, 360p. (1963).

FLEXER, A. S.: Bipolar incompatibility in *Polyporus palustris*. Ph. D. Thesis. Cambridge, Mass.: Harvard University 1963.

FRIES, N.: Researches into the multipolar sexuality of Cyathus striatus PERS. Symbolae Botan. Upsalienses **4**, 1—39 (1940).

— Über das Vorkommen von geographischen Rassen bei *Crucibulum vulgare*. Arch. Mikrobiol. **13**, 182—190 (1943).

—, and L. JONASSON: Über die Infertilität verschiedener Stämme von *Polyporus abietinus* (DICKS.) FR. Svensk. Bot. Tidskr. **35**, 177—193 (1941).

HOROWITZ, N.H., M. FLING, H. MACLEOD, and N. SUEOKA: A genetic study of new structural forms of tyrosinase in *Neurospora*. Genetics **46**, 1015—1024 (1961).

Klushnikova, E. S.: The wild *Psalliota campestris*, its sexual characters and its relation to the cultivated mushroom. Bull. Soc. Vat. Moscou, **48**, 53—58 (In Russian, English summary) (1939).

Kühner, R.: La Genre Mycena. Encyclopedie Mycologique, Vol. 10, Paris (1938).

Lamoure, D.: Hétérocaryose chez les Basidiomycètes amphithalles. Compt. Rend. Acad. Sci. (Paris) **244**, 2841—2843 (1957).

Lange, M.: Species concept in the genus *Coprinus*, a study on the significance of interfertility. Dansk. Bot. Ark. **14**, 1—164 (1952).

Lewis, D.: Comparative incompatibility in angiosperms and fungi. Advan. Genet. **6**, 235—285 (1954).

— Growth and genetics of higher fungi. Nature **190**, 399—400 (1961).

Middleton, R. B.: Sexual and somatic recombination in common-AB heterokaryons of *Schizophyllum commune*. Genetics **50**, 701—710 (1964).

Parag, Y.: Genetic studies on somatic recombination and common-B heterokaryosis in *Schizophyllum commune*. Ph. D. Thesis, Harvard University, Cambridge, Mass. 172p. (1960).

— Mutations in the B incompatibility factor of *Schizophyllum commune*. Proc. Natl. Acad. Sci. (U. S.) **48**, 743—750 (1962).

—, and J. R. Raper: Genetic recombination in a common-B cross of *Schizophyllum commune*. Nature **188**, 765—766 (1961).

Pontecorvo, G.: Trends in Genetic Analysis. New York: Columbia University Press, 145p. (1958).

Prévost, G.: Étude génétique d'un Basidiomycète: *"Coprinus radiatus"* Fr. ex Bolt. Ph. D. Thesis, Paris. 195p. (1962).

Quintanilha, A., and J. Pinto-Lopes: Apercu sur l'état actuel de nos connaissances concernant la »conduite sexuelle« des espèces d'Hymenomycètes. I. Bol. Soc. Broter. 2a ser., **24**, 115—290 (1950).

Raper, C. A., and J. R. Raper: Mutations affecting heterokaryosis in *Schizophyllum commune*. Am. J. Botany **51**, 503—512 (1964).

Raper, J. R.: Tetrapolar sexuality. Quart. Rev. Biol. **28**, 233—259 (1953).

— Sexual versatility and evolutionary processes in fungi. Mycologia **51**, 107—125 (1959a).

— *Schizophyllum umbrinum* Berkeley in culture. Mycologia **51**, 474—476 (1959b).

— The control of sex in fungi. Am. J. Botany **47**, 794—808 (1960a).

— Tetrapolarity in *Schizophyllum fasciatum*. Mycologia **52**, 334—336 (1960b).

—, M. G. Baxter, and A. H. Ellingboe: The genetic structure of the incompatibility factors in *Schizophyllum commune*. The A factor. Proc. Natl. Acad. Sci. (U. S.) **46**, 883—842 (1960).

— —, and R. B. Middleton: The genetic structure of the incompatibility factors in *Schizophyllum commune*. Proc. Natl. Acad. Sci. (U. S.) **44**, 889—900 (1958a).

—, D. H. Boyd, and C. A. Raper: Primary and secondary mutations in the incompatibility loci of *Schizophyllum*. Proc. Natl. Acad. Sci. (U. S.) **53**, 1324—1332 (1965).

—, G. S. Krongelb, and M. G. Baxter: The number and distribution of incompatibility factors in *Schizophyllum*. Am. Naturalist **92**, 221—232 (1958b).

Sarazin, A.: The Cultivated Mushrooms. Morphology, Cytology, Evolution, Nutrition and Cultivation. (English translation.) Mushroom Growers' Association, 45 Bedford Sq., London. 75p. (1955).

SASS, J.E.: The cytological basis for homothallism and heterothallism in the Agaricaceae. Am. J. Botany 16, 663—701 (1929).

SMITH, A.H.: Investigation of two-spored forms in the genus *Mycena*. Mycologia 26, 305—331 (1934).

TAKAHASHI, T.: Complementary genes controlling homothallism in *Saccharomyces*. Genetics 43, 705—715 (1958).

WHITEHOUSE, H.L.K.: Multiple-allelomorph heterothallism in the fungi. New Phytologist 48, 212—244 (1949).

WINGE, Ö., and C. ROBERTS: A gene for diploidization in yeast. Compt. Rend. Trav. Lab. Carlsberg, Serie Physiologique 24, 341—346 (1949).

WOODWARD, D.O.: A gene concept based on genetic and chemical studies in *Neurospora*. Quart. Rev. Biol. 35, 313—323 (1960).

—, C.W.H. PARTRIDGE, and N.H. Giles: Complementation at the ad-4 locus in *Neurospora crassa*. Proc. Natl. Acad. Sci. (U.S.) 44, 1237—1244 (1958).

YANOFSKY, C.: The tryptophan synthetase system. Bacteriol. Rev. 24, 221—245 (1960).

—, and P. St. LAWRENCE: Gene action. Ann. Rev. Microbiol. 14, 311—340 (1960).

ZATTLER, F.: Vererbungsstudien an Hutpilzen (Basidiomyzeten). Z. Botan. 16, 433—499 (1924).

Discussion

Bistis: On the basis of one scheme which has been proposed, bipolars have originated from tetrapolars. If this were so, would you expect to find two groups of bipolars, one derived from the *A* factor and one derived from the *B* factor?

Parag: That is a possibility.

Snider: In another hypothesis for the origin of tetrapolarity, bipolarity may evolve into tetrapolarity by linear duplication, i.e., unequal crossing-over, of a locus. The new locus could subsequently be free to diverge evolutionarily to a new function. Eventually its function might be very different from that of the first locus. Nevertheless, until it did, there would be an overlap of function between the two loci.

The natural history of recombination systems

by

J.H. BURNETT

The control of recombination is one of the most important parameters which determine the genetic composition, and hence the evolutionary potentialities, of any population. A good deal is now known about the mechanisms which bring about recombination in fungi, viz. heterokaryosis, parasexuality and the various mating systems, particularly the incompatibility systems. In 1942 MATHER showed clearly that heterothallism can be regarded as a control system for outbreeding in fungi. He, WHITEHOUSE (1949) and

LEWIS (1954) have all demonstrated the potential degree of outbreeding which can result from the operation of various mating systems. Many other investigators, including KNIEP, VANDENDRIES, BULLER and most of today's speakers, have investigated, and are still investigating, the physiology and genetics of such systems. I am concerned, however, to try to describe how these systems actually operate in natural populations. When considering this question, it is not only necessary to know about the nature of the underlaying mechanisms but also the size of the breeding population, the numbers of mating-type factors and their distribution in space and time. When discussing mating systems, I shall use terms which I first suggested in 1956, especially dimixis and diaphoromixis, since it seems to me that an important distinction can be drawn between those outbreeding forms where there are only two possible kinds of conjugant partners and those (exclusively in Basidiomycetes) where there are more than two possible conjugant partners, whether or not the mating system is controlled by one locus (bipolar) or by two loci (tetrapolar).

In dimictic forms (with two mating-types), a notable feature is that inter-specific crosses are frequently fertile or partially so, whereas, in diaphoromictic forms (with multiple mating-types), there is no unequivocal example of a succesful cross. Of course, so few investigations have been made that it is not yet clear how often the products of inter-specific crosses are viable, but undoubted examples are known in *Allomyces*, yeasts, *Neurospora*, *Cochliobolus* and smut fungi (e. g., WINGE, 1941; NELSON, 1963). On the other hand, it is not completely certain that inter-specific crosses always fail between diaphoromictic forms. Nevertheless, at the present time, this generalisation is not unreasonable. Thus, recombination can occur between more widely differing individuals if they posses dimictic mating systems than if they are diaphoromictic. The significance of this difference is not at all clear at present nor is any kind of explanation available.

Although I have suggested that a very wide degree of outbreeding is possible between dimictic forms, attention must now be drawn to important restrictions which may be imposed on such species in their normal environment. I will take two examples, the genus *Phytophthora* and the Mucorales. It seems probable that the whole problem of sexuality in the biflagellate Oomycetes is likely to be re-interpreted if the recent findings of SANSOME (1963), that the vegetative phase of *Pythium*, *Phytophthora* and *Achlya* is diploid, proves to be the rule for the group. Even if this is the case, however, it will still be true that inter-specific hybridization can readily be achieved between many species of *Phytophthora* (e. g. SAVAGE and CLAYTON, 1962) and that in *P. infestans* and a number of other species, compatibility is based upon a dimictic system (SMOOT et al., 1958; GALINDO and GALLEGLY, 1960). However, as SMOOT et al. have shown very clearly, the dimictic system is confined mainly to the region of Central Mexico. In this region, where numerous tuber-bearing species of *Solanum* occur, both mating-types, *A1* and *A2*, of the potato blight fungus are found. This may also, presumably, be the case in South America where the other important centre of diversity of tuberous Solanums occurs (NIEDERHAUSER, 1956). However, outside these regions, in the U.S.A. and Canada, the West Indies,

Western Europe and South America, only a single mating-type, *A2*, is found. In *P. infestans* in these regions, which include the principal potato

Table 1. *Distribution of mating-types amongst isolations of Mucoraceous fungi.*
(From BLAKESLEE et al., 1927)

Organism	+ mating-type	− mating-type	Neutral
Absidia blakesleeana	19	18	3
caerulea	4	13	5
cylindrospora	1	1	0
dubia	4	20	0
glauca	4	6	0
ramosa	2	1	0
repens	8	1	0
sp. (whorled)	14	18	2
Blakeslea trispora	1	1	0
Chaetocladium brefeldii	?	?	0
Choanephora sp. A	1	1	0
cucurbitarum	5	28	0
Circinella spinosa	34	14	7
umbellata	1	1	0
Cunninghamella sp. A	22	29	2
bertholletiae	12	69	8
echinulata	10	8	0
elegans	25	16	1
Helicostylum piriforme	6	3	0
Mucor griseo-cyanus	3	6	0
hiemalis	1	1	0
dispersus	1	1	0
mucedo	14	11	14
sp. N.	1	1	0
sp. III	3	2	0
sp. IV	6	13	0
sp. V	3	2	0
sp. VI	1	1	0
sp. VII	1	5	0
sp. VIII	1	1	0
Parasitella simplex	1	3	0
Phycomyces blakesleeanus	11	1	3
Rhizopus nigricans	89	62	85
Syncephalastrum racemosum	37	39	4
Other races of above species	7	9	2
Races of species not listed	40	64	99
Totals	393	470	235

growing regions of the World, a sexual outbreeding system does not exist and such variation as the parasite shows, must result either from a change in the mating system, e. g., to homomixis (homothallism); or heterokaryosis, via mutation; or through some form of cytoplasmic heredity (JINKS and GRINDLE, 1963a); or conceivably, through some form of somatic recombination which might have evolved in forms of the fungus outside the central Mexican Region, although there is certainly no evidence for this. It appears, therefore, that variability over the greater part of the range of this species is due to mutation alone rather than to the consequences of recombination.

The Mucorales are the group in which the original work on the sexuality of the fungi was done. The work of BLAKESLEE, BURGEFF, LING YONG, and KÖHLER has shown that dimixis is the rule in outbreeding species and that a wide range of potential inter-specific hybrids can be induced. It is also clear that heterokaryosis via mutation and possibly through hyphal fusion can occur (KÖHLER, 1935). With this high degree of potential outbreeding, it might be expected that these fungi would be highly variable. A cursory study of the taxonomic treatments of the group shows that this expectation is fully justified; they are indeed highly variable. But does this variability really arise as a result of the outbreeding system? Students of soil fungi have repeatedly isolated strains of particular species which turned out to be of the same mating-type, and BLAKESLEE and others showed clearly that the mating-types in any species were often unequally distributed and that 'neutral' strains were by no means uncommon (Table 1). Moreover, it is well-known that zygotes are of very infrequent occurence in natural habitats, except in in-breeding (homothallic) forms. Even when zygotes are found or are produced in culture, it has been the despair of investigators to get them to germinate. In natural environments, therefore, it looks as if the mating system may not operate very efficiently either because of the lack of the two necessary mating-types or through the widespread occurrence of neutral strains. Even when sexual reproduction has taken place, failure of the zygote to germinate means that it is doubtful whether progeny are ever released and re-incorporated in the natural populations. Incidentally, a similar difficulty may occur in a number of water moulds, since here too the germination of the zygotes has rarely been achieved in experiments (e. g., EMERSON, 1950).

In these examples, the potential for outbreeding is great and transcends the limits of taxonomic species or, in the Mucorales, of genera and families. Yet, in nature, it looks as if recombination at any level, down to the intra-specific, may be greatly restricted, either because of the uneven distribution of compatible mating types, or through the lack of progeny resulting from failure of the zygote to germinate.

By contrast, in other fungi amongst the Ascomycetes and Basidiomycetes, obligatory restrictions may be imposed upon the size of the breeding unit. Work at Birmingham (GRINDLE, 1963a, b) has shown clearly that a species such as *Aspergillus nidulans* is divided into a number of groups between which heterokaryons cannot be formed. Recombination, whether mitotic or meiotic, can only take place, therefore, within these groups: the members of different groups are usually identical or extremely similar in their morphology in culture and often occur in the same locality. It is not, of course, possible to make a complete genetical analysis of this situation, but there is suggestive evidence that control is exercised through nuclear genes (JINKS and GRINDLE, 1963b). A somewhat similar situation has been analysed in great detail by ESSER in *Podospora anserina* and is described above,

Table 2. *Restrictions on mating ability between groups within recognized taxonomic species in relation to differences in mating system, morphology and ecology*

Organism	Differences in			Inter-Sterility	Author
	Ma-ting Syst.	Mor-pho-logy	Eco-logy		
Mycocalia denudata and M. castanea	+	+	+	Complete	BURNETT & BOULTER, 1963
Corticium coronilla	+	+	?	Complete	BRIGGS, 1937
Gloeocystidium tenue	+	+	?	Complete	BOIDIN, 1951
Peniophora spp.	—	+	+	Complete or partial	McKEEN, 1952
Fomes ignarius	—	+	+	Complete	VERRALL, 1937
Auricularia auricula-judae	—	+	+	Partial	BARNETT, 1937
Polyporus abietinus & forms	—	+	±	Partial or none	MACRAE, 1941; RAESTED, 1941
Gloeocystidium tenue	—	+	?	Complete	BOIDIN, 1951
Coprinus micaceus	—	+	?	Partial	KÜHNER, ROMAGNESI & YEN, 1947
Fomes pinicola	—	—	+	Partial	MOUNCE, 1929
Fomes pinicola	—	—	—	Complete	MOUNCE & MACRAE, 1938
Polyporus betulinus	—	—	—	Complete	BURNETT, unpublished
Mycocalia denudata	—	—	—	Complete	BURNETT & BOULTER, 1963
Coprinus macrorhizus f. microsporus	—	?	?	Complete	KIMURA, 1952
Coprinus callinus	—	—	?	Complete	LANGE, 1952
Coprinus subimpatiens	—	—	?	Complete	LANGE, 1952

pp. 6–13. All that need be said here is that he has provided firm evidence that this species is divided up into a number of inter-sterile or partially inter-fertile groups which are determined by the segregation and recombination of four pairs of allelomorphs. In his case, however, both heterokaryosis and mating ability are determined by these genes.

It is somewhat curious that the occurrence of these groups with re-stricted mating ability should occur in species in which some degree of inbreeding already exists, for *A.nidulans* is a homomictic (homothallic) fungus and *P. anserina* is a homoheteromictic (secondarily homothallic) fungus. It seems probable that the occurence of this phenomenon in inbreeding Ascomycetes is not the rule but merely reflects the very small number of fungi which have been investigated. In the Basidiomycetes, at least, groups with restricted mating ability occur in normal, outbreeding species. Several examples of this phenomenon have been described and are illustrated in Table 2; all the fungi shown in this table have multiple mating-types.

In some cases there is partial fertility between the groups, and this may, or may not, be associated with differences in morphology or even differences in the host or substrate. Examples of such situations occur in *Fomes pinicola* (Fig. 1). In *Fomes pinicola* there is morphological variation amongst sporophores and isolations but this is unrelated to the three inter-sterile

Figure 1. The distribution of the three breeding groups in *Fomes pinicola*
(from Mounce, I. and R. Macrae, 1938)

groups. Groups A and B, apparently confined to N. America, may even be adjacent on the same tree but they are totally inter-sterile. However, fusions and dikaryotic hyphae can be obtained between isolates either from Group A (readily) or Group B (very rarely) with Group C, whose members lie outside N. America (Mounce and Macrae, 1938). *Auricularia auricula-judae* is an example where there are few morphological differences but in which host specialisation is associated with mating ability. In the U.S.A., isolates from conifers are only partially fertile with those from deciduous trees (Barnett, 1937). Crosses were not made on a wider scale for, in Europe, this fungus is largely restricted to elder (*Sambucus nigra*). The case of *Polyporus abietinus* is more complex. In N. America, poroid, irpicoid, and an intermediate 'lamellate' form occur on conifers and show no host specialization (Macrae, 1941). In Europe a poroid form occurs, restricted largely to *Pinus*, and in

Norway, at least, there also occurs an irpicoid form on conifers generally which is often recognized as a separate species, *Irpex fusco-violaceus* (RAESTAD, 1941). The inter-crossing relationships are shown in Fig. 2. The behaviour of the poroid forms resembles that of the groups within *F. pinicola*, while the irpicoid forms co-exist as a separate intra-fertile group. The most surprising feature is that the N. American, bipolar, 'lamellate' form is apparently partially fertile with the tetrapolar, European, poroid form. This is a unique situation which demands re-investigation (but see discussion, p. 113).

In other species, however, completely inter-sterile groups occur between isolates which show no constant differences in their morphology, ecology or mating systems. Such examples occur both in bipolar and tetrapolar fungi. Perhaps the most striking example of this type occurs in *Mycocalia denudata*. Two fruiting bodies of this species that yielded inter-sterile progeny, 144 and 144N, were obtained from what was thought to be a

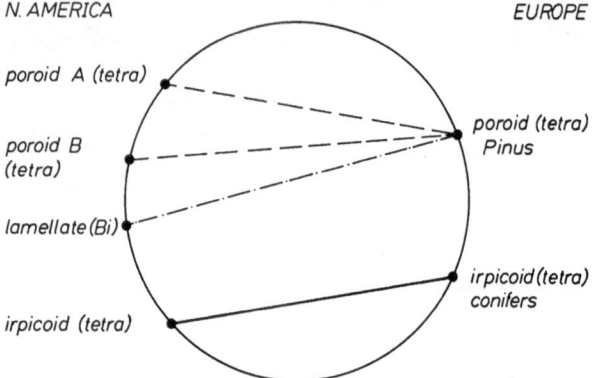

Figure 2. Interbreeding relationships between the different breeding groups of *Polyporus abietinus*. —— fully inter-fertile; — — — partially inter-fertile; —·—·— rarely inter-fertile. The presence of no connecting lines across the diagram between groups indicates total inter-sterility. (Based on data of MACRAE, R., 1941 and R. RAESTAD, 1941)

group of peridiola derived from a single individual. In fact, however, they had clearly been obtained from two individuals whose fruiting bodies had become confluent. The growth rate and colour of the mycelia, the colour and size range of the peridiola, and the ranges of basidiospore sizes of the two isolates were compared, and no constant differences could be found between them, although they clearly shared the same ecological niche! Because the groups within such species in Table 2 are totally inter-sterile, it has not been possible to analyse the genetical situation and, in some cases, e. g., *Mycocalia denudata*, even hyphal fusions failed to occur between the inter-sterile groups.

It is convenient here to point out that inbreeding may also be imposed by cytological or morphological means. *M. denudata* is an example; here a gene, *pd*, is known which causes an additional division to occur in the basidium after meiosis to yield an 8-nucleate condition. All the nuclei migrate into the four basidiospores, so that these may be either homokaryotic or heterokaryotic for mating-type. Hence, *pd* is responsible for some degree of facultative inbreeding in this fungus (BURNETT and BOULTER, 1963). Since 8-nucleate basidia are not uncommon in agarics (KÜHNER, 1945), this phenomenon may be quite widespread in such fungi. A morphological modification which brings about the same result is the production

Table 3. *Numbers of mating-type (incompatibility) factors identified and estimated total numbers based on these finds*

BIPOLAR SPECIES

	Mono-karyons examined	Alleles identified	Estimated total alleles	Author
Auricularia auricula-judae	10	10	v. large	BARNETT, 1937
Coprinus comatus	11	9	30	SAUNDERS unpublished
	22	11	15	BRUNSWIK, 1924
Fomes roseus	10	9	50	MOUNCE & MACRAE, 1937
Fomes subroseus	20	20	v. large	MOUNCE & MACRAE, 1937
Polyporus betulinus	201	28	30	SAUNDERS, 1956: BURNETT, unpublished.
Mycocalia denudata I	6	4	7	BURNETT &
II	14	7	9	BOULTER, 1963

TETRAPOLAR SPECIES

		A	B	A	B	
Coprinus fimetarius	27	27	27	v. large		BRUNSWIK, 1924
Coprinus lagopus	14	14	14	v. large		HANNA, 1925
Coprinus macrorhizus	20	19	19	240	240	KIMURA, 1952
Polyporus abietinus	28	23	26	100	200	FRIES & JONASSON, 1941
Polyporus obtusus	48	39	39	85	85	EGGERTSON 1953
Polystictus versicolor	46	20	20	25	25	PARTINGTON, 1959
Schizophyllum commune	114	96	56	339	64	RAPER *et al.* 1958
Crucibulum vulgare	30	3	11	5	15	FRIES, 1936
Cyathus striatus	18	4	5	5	5	FRIES, 1936, 1940

of 2 or 3 sterigmata and hence that number of basidiospores, on a basidium with normal meiosis in which all the nuclei migrate into the spores. In some cases, the morphology and migration pattern is fortuitous, e. g., *C. plagioporus* (LAMOURE, 1951), in others it is genetically fixed and some unknown (cytological?) mechanism ensures that the vast majority of spores are heterokaryotic for mating-type factors, e. g., *Coprinus sassii* (Lange, 1952). An analogous situation occurs in certain Ascomycetes where the normal 8-spored ascus is replaced by a 4-spored one, each spore carrying both mating-type factors as a consequence of appropriate genetical and/or cytological mechanisms, e. g., *Neurospora tetrasperma* (Shear and Dodge, 1927; Dodge et al., 1950), *Podospora anserina* (Dowding, 1931; Franke, 1957, 1962).

These restrictions on outbreeding are in great contrast to the situation in a fungus such as *Schizophyllum commune*, in which isolates from all over the world are completely compatible with each other (Raper et al., 1958). In fungi such as this, and even within some of the restricted breeding groups which have been described earlier, e. g., European poroid *P. abietinus*, the

Table 4, *The potential outbreeding efficiency of small, relatively isolated, local populations of various tetrapolar fungi*

Organism	Potential outbreeding efficiency	% potential outbreeding efficiency	Area or distance between populations	Author
Schizophyllum commune	3.96	98.5	10 acres	Roshal, 1950
Polyporus obtusus	3.89	96.9	?	Eggertson, 1953
Polystictus versicolor	3.68	91.6	10 miles	Partington, 1959
P. versicolor	3.24	80.5	Single stump	Partington, 1959

potential outbreeding is very high indeed (Fries and Jonasson, 1941), because the mating system is based upon large numbers of mating-type or incompatibility factors; actual numbers found and estimated total numbers are shown in Table 3.

Most of the data in Table 3 are based upon isolates taken from a wide range of sources and only a limited number of studies have been made upon populations from limited regions. The most detailed investigations are those of Roshal (1950), Eggertson (1953) and my own group. Some data are set out in Table 4. In each case, the potential outbreeding bias has been calculated. This is the ratio of (potential) non-sister matings to (potential) sister matings. Its value varies with the mating system, tetrapolar or bipolar, and the number of allelomorphic mating-type factors as shown in Fig. 3. It can be seen from this figure that, whatever the mating system, the value of the outbreeding bias becomes asymptotic once 20—30 alleles are available. Thus, of the tetrapolar species studied, only the population of *P. versicolor* restricted to an isolated stump shows any serious impairment of outbreeding potentialities compared with the other populations. Even here, potentially 81% of the maximum outbreeding bias is available (3.24 : 4.00).

It is unfortunate that so little is known of the quantitative effects on populations of variations in the outbreeding potential. While the difference between 10 and 25 mating-type factors, as in the populations of *P. versicolor*, represents a difference of 11% in outbreeding bias, that between the larger of these populations and *S. commune*, for example, represents an increase of only a further 8%, yet the difference in the number of mating-type factors is 339A and 64B in *S. commune* to 25A and 25B in *P. versicolor*. Teleologically speaking, it seems as though the number of mating-type factors in some tetrapolar fungi is unnecessarily high in order to ensure effective outbreeding! In bipolar fungi, the estimates for numbers of allelomorphs in

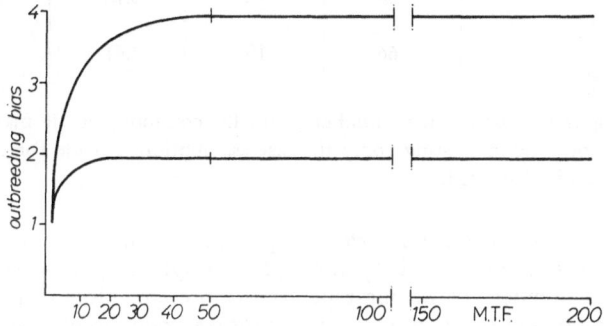

Figure 3. Diagram to illustrate the relationship between the number of mating-type factors (M.T.F.) and the outbreeding bias for tetrapolar, (upper curve) and bipolar (lower curve) diaphoromictic fungi

widespread populations is usually much lower (Table 3) and yet large enough, e. g., 30 in *P. betulinus*, to ensure an adequate outbreeding bias. In the gasteromycetes the situation is obscure and has been discussed elsewhere (BURNETT and BOULTER, 1963).

Two points may be made concerning these estimates. Firstly, the calculations are based upon the assumption that all mating-type factors are equally common in the population. This certainly seems to hold for the world-wide sample of *S. commune* (RAPER, et al., 1958) and for the U.K. sample of *P. betulinus* (SAUNDERS, 1956). In small isolated populations of this latter fungus, however, this may not be the case. The Lancashire coast (Freshfield) and Scottish (Stravithie) populations were each separated by at least 1 kilometre from other populations (the fungus is confined to *Betula* spp. which makes such estimates possible and reliable) and here there is clearly a departure from equality of mating-type factors. The data for these populations compared with those for the species as a whole are shown in Table 5, where it can be seen that there is a considerable reduction in the outbreeding bias of these small populations.

The second difficulty, inherent in calculations such as these, is that it is not easy to measure the degree of reproductive isolation or the effective breeding size of a local population. There is virtually no evidence of the effectiveness of anemophilous dispersal, although the data of GREGORY

(1961) suggest that only 1% of fungal propagules are dispersed beyond 100 m. However, even a single basidiospore every few years is sufficient to increase the outbreeding efficiency of a population with a perennial mycelium. This latter point also affects the issue. Not every mycelium fruits

Table 5. *Potential outbreeding in populations of P. betulinus*

Population	Mono-karyons sampled	Mating-type factors detected	Outbreeding bias	% Outbreeding
British sample	201	28	3.86*	96.4*
Stravithie (Scotland)	58	16	2.78	76.4
Freshfield (England)	66	19	3.61	78.3

* These figures assume equal numbers, equally common, of all mating-type factors, and random mating; only the last assumption is made for the other two populations studied.

every year, as the data for *P. betulinus* sporophores produced on a single tree in successive years (Table 6) shows. Thus single samplings may underestimate the numbers of mating-type factors present. Moreover, Nobles (1958) has pointed out that apparently monokaryotic mycelia are not uncommon in the host tissue amongst wood-destroying polypores, and this is

Table 6. *Mating-type constitution of fruit bodies of P. betulinus on the same trunk in successive years at Stravithie*

Year	No. of sporophores	Mating-types from the base to apex
1955	8	3.8 : 3.12 : 3.8 : 4.20 : 4.20 : 9.10 : 3.8 : 4.20
1956	5	3.8 : 3.4 : 3.4 : 4.20 : 8.9
1957	7	3.8 : 3.12 : 3.4 : 4.20 : 9.10 : 8.9 : 3.8
1958	6	3.12 : 3.4 : 3.8 : 4.20 : 4.20 : 4.20
1959	6	3.8 : 3.8 : 3.12 : 3.4 : 9.10 : 3.9
1960	8	3.8 : 3.12 : 3.8 : 3.4 : 4.20 : 4.9 : 4.20 : 4.13.

also true of soil isolations of Basidiomycetes. Over many years, therefore, compatible fusions may occur and hence sporophore production which adds new mating-type factors to the breeding population.

These considerations lead me to my final topic, the nature and significance of mycelium. In our work with *P. versicolor*, *P. betulinus*, *Coprinus comatus* and other fungi, we have noticed that sporophores in close proximity often carry the same mating-type factors but in different combinations (Fig. 4). We, and others, have shown that there is little restriction on hyphal anastomoses: monokaryons, whether genetically alike or different, can fuse with monokaryons or with dikaryons and similarly dikaryons with dikaryons. We have suggested, therefore, that the mycelium exists in its natural environment as a single physiological and ecological unit although genetically a mosaic (BURNETT and PARTINGTON, 1957). Now, if this were indeed

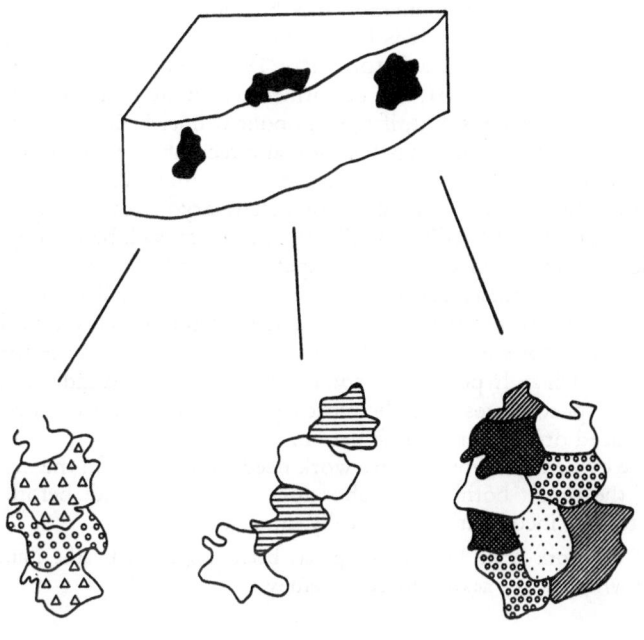

| A 1.2 | A 1.6 | A 2.4 | A 3.4 | A 4.6 | A 5.6 | constitution |
| B 1.2 | B 1.6 | B 2.4 | B 3.4 | B 4.6 | B 5.6 | unknown |

Figure 4. Diagram to illustrate the distribution of sporophores of *Polystictus versicolor* on a stump. a) General distribution to show 3 groups of sporophores each 30 cm apart. b) Diagram to illustrate mating type constitution of sporophores of each group

the case, it might be possible for nuclear migration and hence association to occur between nuclei from different parts of the mycelium. There is clear evidence of nuclear migration in basidiomycete mycelium, in at least, three species, viz. *S. commune* (SNIDER, 1963), *C. lagopus* (SWIEZYNSKI and DAY, 1960b) and *P. versicolor* (PARTINGTON, 1959). As a consequence of such studies, it is known that certain restrictions may be imposed on nuclear migration and hence association. On the other hand, it is also known that diploid nuclei can arise in the mycelium and that, by various means, somatic recombination can take place. Drs. SNIDER, ELLINGBOE and PRUD'HOMME will deal with these problems in great detail later (see pp. 36–52). All that concerns me here is the significance of these facts for mycelia in natural habitats.

If such behaviour is widespread, and this is not known, the occurrence of perennial, apparently sterile mycelia takes on added significance. Recombinant nuclei may be generated in different parts of an heterokaryotic mycelium through somatic recombination and thus new, effective variants can arise to exploit the substrate. It is known both in *C. lagopus* (SWIEZYNSKI and DAY, 1960a) and *S. commune* (MIDDLETON, 1964) that *common AB* heterokaryons, i. e., *A1 B1 + A1 B1*, can be formed which are indistinguishable from normal monokaryons, but, if the component nuclei differ genetically at loci other than that for mating-type, somatic recombinations can occur. It is not inconceivable that some of the apparent monokaryons isolated from nature, and referred to earlier, are heterokaryons of this type. In summary, I suggest that 'sterile' mycelia in nature may well be undergoing l.c.-somatic recombination and thus be capable of producing new mycelial variants. This is, of course, a speculation. Yet it does rationalize the observation known to any mycologist who isolates Basidiomycetes, especially from wood, soil, root surfaces etc. that many are apparently sterile forms. A corollary to such an hypothesis, if somatic recombination is indeed widespread, is that sporophores could be more significant as agents of dispersal than as localized organs for recombination.

It will be evident that much more work needs to be done on the mating systems of the fungi, both in the laboratory and in the field, before the natural history of incompatibility is really understood.

Most of the work done by my group has been supported by the Nuffield Foundation whose help I acknowledge gratefully.

References

BARNETT, H. L.: Studies on the sexuality of the Heterobasideae. Mycologia **29**, 626—649 (1937).

BIGGS, R.: The species concept in *Corticium coronilla*. Mycologia **29**, 686—706 (1937).

BLAKESLEE, A. F., J. L. CARTLEDGE, D. S. WELCH, and A. D. BERGNER: Sexual dimorphism in Mucorales 1. Intraspecific reactions. Botan. Gaz. **84**, 27—90 (1927).

BOIDIN, J.: Sur l'existence de races interstériles chez *Gloeocystidium tenue* (PAT); Étude morphologique et comportement nucléaire de leurs cultures. Bull. Soc. Mycol. France **66**, 204—219 (1950).

BRUNSWIK, H.: Untersuchungen über die Geschlechts- und Kernverhältnisse bei der Hymenomyzeten-Gattung *Coprinus*. Bot. Abh. K. Goebel **5**, 1—152 (1924).

BURNETT, J. H.: The mating systems of fungi I. New Phytologist **55**, 50—90 (1956).

—, and M. E. BOULTER: The mating systems of fungi II. Mating systems of the Gasteromycetes *Mycocalia denudata* and *M. duriaeana*. New Phytologist **62**, 217—236 (1963).

—, and D. PARTINGTON: Spatial distribution of fungal mating-type factors. Proc. Roy. Phys. Soc. Edinburgh **26**, 61—68 (1957).

DODGE, B.O., J.R. SINGLETON, and A. ROLNIK: Studies on lethal E gene in *Neurospora tetrasperma*, including chromosome counts also in races of N. sitophila. Proc. Am. Phil. Soc. **94**, 38—52 (1950).

DOWDING, E.S.: The sexuality of the normal, giant and dwarf spores of *Pleurage anserina* (CES.) KUNTZE. Ann. Botany **45**, 1—14 (1931).

EGGERTSON, E.: An estimate of the number of alleles at the loci for heterothallism in a local concentration of *Polyporus obtusus*. Canad. J. Botany **31**, 710—759 (1953).

EMERSON, R.: Current trends of experimental research in aquatic Phycomycetes. Ann. Rev. Microbiol. **4**, 169—200 (1950).

FRANKE, G.: Die Cytologie der Ascusentwicklung von *Podospora anserina*. Z. indukt. Abstamm.- u. Vererb.-L. **88**, 159—160 (1957).

— Versuche zur Genomverdoppelung des Ascomyceten *Podospora anserina*. Z. Vererb.-L. **93**, 109—117 (1962).

FRIES, N.: *Crucibulum vulgare* TUL. und *Cyathus striatus* PERS. zwei Gasteromyceten mit tetrapolarer Geschlechts-Verteilung. Botan. Notiser **567**—574 (1936).

— Researches into the multipolar sexuality of *Cyathus striatus* PERS. Symbolae Botan. Upsalienses **4**, 5—39 (1940).

—, and L. JONASSON: Über die Infertilität verschiedener Stämme von *Polyporus abietinus* (DICKS.) FR. Svensk. Bot. Tidskr. **35**, 177—193 (1941).

GALINDO, J., and M.E. GALLEGLY: The nature of sexuality in *Phytophthora infestans*. Phytopathology **50**, 123—128 (1960).

GREGORY, P.H.: Microbiology of the Atmosphere, London (1961).

GRINDLE, M.: Heterokaryon incompatibility of unrelated strains in the *Aspergillus nidulans* group. Heredity **18**, 191—204 (1963a).

— Heterokaryon compatability of closely related wild isolates of *Aspergillus nidulans*. Heredity **18**, 397—405 (1963b).

HANNA, W.F.: The problem of sex in *Coprinus lagopus*. Ann. Botany **39**, 431—457 (1925).

JINKS, J.L., and M. GRINDLE: Changes induced by training in *Phytophthora infestans*. Heredity **18**, 245—264 (1963a).

— — The genetical basis of heterokaryon incompatibility in *Aspergillus nidulans*. Heredity **18**, 407—411 (1963b).

KIMURA, K.: Diploidization in the Hymenomycetes I. Preliminary experiments. Biol. J. Okayama Univ. **1**, 72—83 (1952).

KÖHLER, F.: Genetische Studien an *Mucor mucedo* (BREFL.) I; II; III. Z. indukt. Abstamm.- u. Vererb.-L. **70**, 1—26; 27—39; 40—54 (1935).

KÜHNER, R.: Le problème de la filiation des Agaricales à la lumière de nouvelles observations d'ordre cytologique sur les Agaricales Leucosporées. Bull. Soc. Linn. Lyon. **14**, 160—164 (1945).

—, H. ROMAGNESI, and M.C. YEN: Différences morphologiques entre plusiers souches de *Coprins* de la section *Micacei* et confrontation de leurs haplontes. Bull. Soc. Mycol. France **63**, 169—186 (1947).

LAMOURE, D.: Hétérocaryose chez les Basidiomycètes amphithalles. Compt. Rend. Acad. Sci. (Paris) **244**, 2841—2843 (1957).

LANGE, M.: Species concept in the genus *Coprinus*. Dansk. bot. Arkiv. **14**, 7—162 (1952).

LEWIS, D.: Comparative incompatibility in Angiosperms and Fungi. Advan. Genet. 6, 235—285 (1954).

MACRAE, R.: Genetical and sexual studies in some higher Basidiomycetes. Ph. D. Thesis, University of Toronto (1941).

MATHER, K.: Heterothally as an outbreeding mechanism in fungi. Nature, London 149, 54—57 (1942).

MCKEEN, C.V.: A cultural and taxonomic study of three species of *Peniophora*. Canad. J. Botany 30, 764—787 (1952).

MIDDLETON, R.B.: Sexual and somatic recombination in common-AB hetero-caryons. Am. J. Botany 51, 379—387 (1964).

MOUNCE, I.: Studies in forest pathology. The biology of *Fomes pinicola* (Sw.) COOKE. Bull. Dept. Agric. Canada 111, 1—77 (1929).

—, and R. MACRAE: The behaviour of paired monosporus mycelia of *Fomes roseus* (ABB. & SCHW.) COOKE and *F. subroseus* (WIER.) OVERH. Canad. J. Res. C, 15, 154—161 (1937).

— — Infertility phenomena in *Fomes pinicola*. Canad. J. Res. C. 16, 364—376 (1938).

NELSON, R.R.: Interspecific hybridization in fungi. Ann. Rev. Microbiol. 17, 31—48 (1963).

NIEDERHAUSER, J.S.: The blight, the blighter and the blighted. Ann. N.Y. Acad. Sci., Ser. II, 19, 55—63 (1956).

NOBLES, M.K.: Cultural characters as a guide to the taxonomy and phylogeny of the Polyporaceae. Canad. J. Botany 36, 883—926 (1958).

PARTINGTON, M.: Mating systems of fungi with special reference to *Polystictus versicolor*. Ph. D. Thesis, University of St. Andrews (1959).

RAESTEAD, R.: The relation between *Polyporus abietinus* (DICKS ex FR.) FR. and *Irpex fusco-violaceus* (EHRENB. ex FR.) FR. Nytt. Mag. Naturvid. 81, 207—231 (1941).

RAPER, J.R., G.S. KRONGELB, and M.G. BAXTER: The number and distribution of incompatibility factors in *Schizophyllum commune*. Am. Naturalist 92, 221—234 (1958).

ROSHAL, J.Y.: Incompatibility factors in a population of *Schizophyllum commune*. Ph. D. Thesis, University of Chicago 1950.

SANSOME, E.R.: Meiosis in Pythium debaryanum Hesse and its significance in the life-history of the Biflagellatae. Trans. Brit. Mycol. Soc. 46, 63—72 (1963.)

SAUNDERS, M.: The distribution of fungal mating-type factors with special reference to *Polyporus betulinus*. M. Sc. Thesis, University of Liverpool 1956.

SAVAGE, E.J., and C.W. CLAYTON: Interspecific hybridization in the genus *Phytophthora*. Phytopathology 52, 1220 (1962).

SHEAR, C.L., and B.O. Dodge: Life histories of the red bread-mould fungi of the *Monilia sitophila* group. J. Agr. Res. 34, 1019—1042 (1927).

SNIDER, P.J.: Genetic evidence for nuclear migration in Basidiomycetes. Genetics 48, 47—55 (1963).

SMOOT, J.J., F.J. GOUGH, H.A. LAMEY, J.J. EICHENMULLER, and M.E. GALLE-GLY: Production and germination of oospores of *Phytophthora infestans* Phytopathology 48, 165—171 (1958).

SWIEZYNSKI, K.M., and P.R. DAY: Heterokaryon formation in *Coprinus lagopus*. Genet. Res. Camb. **1**, 114—128 (1960a).

— — Migration of nuclei in *Coprinus lagopus*. Genet. Res. Camb. **1**, 129—139 (1960b).

VERRALL, A.F.: Variation in *Fomes ignarius*. Tech. Bull. Minn. Agr. Expt. Sta. **117**, 1—41 (1937).

WHITEHOUSE, H.L.K.: Heterothallism and sex in fungi. Biol. Rev. **24**, 411—447 (1949).

WINGE, Ö.: Le croissement inter-specifique chez les champignons. Scientia Genetica **2**, 171—189 (1941).

Discussion

Weresub: Dr. MACRAE has been doing further work with *Polyporus abietinus* and the picture is quite different there. No pairing occurs between the lamellate species of North America and the poroid species of Europe. This seems to be more consistent with what you have said than what had previously been reported.

Burnett: That's what I would have expected.

Snider: I'd like to know if there are any established examples in the higher Basidiomycetes of fruit bodies from nature which contain more than two kinds of nuclei.

Burnett: Not to my knowledge.

Snider: Then a single fruit body, in general, arises from two nuclei?

Burnett: Yes, I think so.

Esser: In Ascomycetes, in *Sordaria*, we have evidence, by using spore color markers, that the original cell that gives rise to the ascogonium can contain at least three nuclei. From this multinucleate ascogonium the nuclei can unite in pairs to give different types of asci.

The genetical interest of incompatibility in fungi

by

KENNETH MATHER

The incompatibility system in fungi is obviously of great interest to the geneticist in its own right. What is its genetical architecture? How does it function? How has it evolved? How efficient is it, and what are the consequences of its breakdown? As this symposium has shown us, answers are emerging to some of these questions. And as they emerge, they are showing us how studies of the incompatibility system can throw light on matters and mechanisms of even broader genetical and biological interest.

Perhaps the most advanced study is that of the structure and action of the incompatibility factors themselves, the A and B factors, which determine, in the immediate sense, the incompatibility properties of the hyphae that bear them. It is clear that each of these factors is characteristically composed of two component loci, a and β, which recombine by crossing-over to give new factors, having new specificities in their action. In *Schizophyllum*, at any rate, the recombination between the component loci that have been recognised is surprisingly high, and one is prompted to ask why it has not been reduced, and the loci bound more closely together in inheritance, by closer regulation of the positioning of crossing-over in the chromosomes or by the association of inversions with the a and β loci. Assuming that recombination as observed in experiment is a fair guide to the recombination that occurs in nature, one must suppose that the flexibility with which the recombination endows the factor is of a selective advantage, though what that advantage can be is by no means yet clear.

The occurrence of this recombination has, however, enabled us to see clearly a distinction which we could only surmise in the incompatibility genes of flowering plants. Lewis' studies of the S gene in *Oenothera organensis* (see review 1954) have revealed its composite nature, yet the only changes he could observe or induce in it were towards loss of function. Such mutations have been seen in fungi too; but here they can be set in contrast with the recombinants which give new functional incompatibility factors. The distinction between the functional reshuffling of operative genes and the degradation change following from loss of operative capacity of these genes is clear. At the same time it still leaves us with the question of how these genes, capable of working together in this system of functional variation by reshuffling, themselves came into being. Perhaps the further study of the fine genetical structure of the A and B factors will reveal this to us.

In another respect the analysis of the genetical control of incompatibility in fungi has barely begun. In higher plants and animals it is now a commonplace that the phases of a polymorphic system (whether expressing itself in incompatibility, in protective mimicry or other ways), though switched by the immediately recognisable genes, owe their functional adjustment and efficiency to the balance which natural selection has produced in the background genotype. That other genes affect the working of incompatibility has been shown in fungi, but we still know little about the action of the background genotype and its importance in the operation and evolution of the incompatibility system as a whole, excellent material though fungi provide for their investigation.

Almost as a by-product, the study of the incompatibility genes has revealed that the nuclei of a basidiomycetous dikaryon can exchange genes before the normal time of fusion and meiosis. Generally, it is clear that groups of genes are involved, but sometimes the observations suggest some process of change or exchange confined to the incompatibility genes themselves. The interest here is, of course, the general one of the mechanisms that must be involved. The parasexual cycle is an obvious mechanism to invoke where groups of genes are recombined. The biological significance of this cycle is, however, far from clear. It is generally represented as an adaptively significant mechanism, offering a means of recombination and variation as an alternative to normal sexual reproduction and therefore of importance especially in imperfect fungi. It may on the other hand be equally well regarded as representing no more than a non-adaptive break-down of normal nuclear behaviour.

The association of two haploid nuclei in the dikaryon of a Basidiomy-cete offers no advantage for the development of the fungus that could not equally be secured by their fusion, yet they are kept apart while maintained in step by the elaborate mechanisms of synchronous division and clamp connections. And the moment they fuse meiosis sets in. This suggests strongly that these fungi have not, so to speak, mastered the problem of managing a diploid nucleus, that they do not possess the refined genetic balance to secure the accurate timing in relation to one another of the cycles of chromosome division and nuclear division necessary for diploid mitosis to follow a normal path, undisturbed by chance partial and uncontrolled pairing of homologues with an irregular mixture of mitotic and quasi-meiotic behaviour as the result. Such an interpretation would accord with the known facts of parasexuality, which can thus be regarded in the Basidiomycetes, at any rate, as no more than the irregular and non-adaptive consequences of a chance nuclear fusion arising from failure of the normal mechanism of the dikaryon to keep the haploid nuclei apart.

However this may be, and whatever the significance of parasexuality, specific change confined to the incompatibility genes themselves and failing to affect even loci lying between them would seem to imply some further mechanism of a different kind — some form of transduction or transfor-mation. If transduction operates occasionally between genes in haploid nuclei held apart in a dikaryon, it might be expected to occur even more frequently between homologous genes in a diploid nucleus. This would on the one hand lead one to expect such changes to be associated especially with the production of basidia, and on the other suggest a further advantage of the dikaryotic over the diploid state for the maintenance of genic inte-grity in these fungi.

Turning next to the biological significance of incompatibility itself, it has long been recognised as a means of promoting outbreeding and there-fore genetical flexibility in the fungi, as in other plants that display it. Now the promotion of outbreeding is but one aspect of the control of the breed-ing system, with all that this implies for fitness and flexibility, as our studies of higher plants have shown us. Some species are inbreeders, not out-breeders, and we must in general regard these forms which regularly inbreed as derived from out-breeding ancestors. It is not surprising there-fore to find inbreeders bearing relics of the ancestral outbreeding mecha-nisms, including cryptic or suppressed incompatibility systems. This situ-ation has long been known in flowering plants and evidently it occurs in fungi too.

Now an inbreeding species requires no further mechanism for isolating it from other species, since its member individuals do not normally cross with one another, let alone with other forms. An out-breeding species on the other hand does require, and as other experience with flowering plants testifies, does normally possess isolating mechanisms for preventing these overwide outcrosses with other species. In the higher plants such isolation can depend on the pollen-style relation just as does incompatibility within the species. In Fungi, too, Professor ESSER's investigations (see this symposium p. 6–13) show us that while in one form incompatibility promotes outbreeding, his heterogenic incompatibility restricts outbreeding. Heterogenic incompatibility is an isolating mechanism, a step on the road to speciation. The function of the isolating mechanism is different from, albeit complementary to, that of incompatibility as we normally use the term. For the one prevents mating between the genetically too-unlike (heterogenic incompatibility) while the other prevents it between the genetically too-like (homogenic incompatibility).

Relatively little is yet known of isolation in fungi. It has, however, been observed to occur in suprising places. In *Aspergillus nidulans*, my former colleague Dr. GRINDLE (1963) has found that the majority of his isolates from the wild will not come together to form the heterokaryons which are pre-requisites for genic recombination by the sexual process. Similar failure to form heterokaryons is known in laboratory stocks of *Neurospora* (GARNJOBST, 1953), though here the failure does not of course prevent sexual reproduction. The isolates of *Aspergillus nidulans* in fact fall into groups within which heterokaryon formation is a normal occurence but between which it is absent or at any rate very rare, the isolates being morphologically more similar within than between groups. To use the Drosophilist's terminology, *A. nidulans* appears to consist of a constellation of sibling species, closely resembling one another in form but virtually isolated from one another genetically. The evidence coming to hand suggests that the formation of heterokaryons is even rarer in the *Aspergillus glaucus* group.

In the absence of heterokaryons, sexual reproduction has no significance at least as far as genic recombination is concerned. (Nor, it may be observed in passing, has parasexuality). Why then does sexual reproduction persist? Why does not the fungus become imperfect? Various answers may be returned to these questions. Other species of *Aspergillus* are imperfect and perhaps *A. nidulans* and *A. glaucus* are on the way. Again, some heterokaryons are formed even though only from infrequent combinations, and it may be that the advantage of recombination even as a relative rarity is sufficient to preserve the mechanism of sexual reproduction. It may be too that sexual reproduction is associated with the formation of resting spores obtained in no other way. Finally we know that, especially in *A. glaucus*, continued asexual propagation, whether by hyphae or conidia, results in the accumulation of cytoplasmic changes bringing about loss of sexuality and finally what JINKS (1956) has termed vegetative death. Cytoplasmically determined senescence is also known in *Podospora* (RIZET, 1957). In *Aspergillus glaucus* a single passage through a sexual spore, irrespective of whether this is from heterokaryon or homokaryon, clears away these and other cytoplasmic changes and restores normal development and vigour (MATHER and JINKS, 1958). The need for this cleaning-up of the cytoplasm could result in the retention of the sexual cycle even after the loss of its function in promoting recombination of the genes. So far as we know, no such cleaning-up is associated with parasexuality, though we have evidence that it can be transferred, at least in respect of some cytoplasmic changes, to conidial production.

One last point. The incompatibility genes of *Schizophyllum* were the subject of the widest geographical investigation yet undertaken in fungi, when RAPER and his collobarators (1958) showed that the factors of the *A* series occuring in the wild should probably be numbered by the hundreds. He also showed that there was no evidence of any deviation from random distribution of the *A* and *B* factors even in widely separated regions of the world. This finding might be interpreted in two ways. It could imply that given the presence of a limited number of alleles of the component loci, no matter in what the combinations they occured, the complete range of alternative factors could be reproduced by recombination — a view which accords well with the high recombination between the two loci of each factor, *A* and *B*. The alternative view is, of course, that the fungus is being dispersed constantly, rapidly and extensively, so that no genetical differences characterize widely separated populations. This is an unattractive interpretation, but it is one that could be tested by comparison of the relational balance, or co-adaptation, of gene combinations derived from the same and from different populations, using characters like vigour of growth as the tell-tale. On this view they should be the same, but if migration is not so constant and ubiquitous, they should be different. Such information has not sofar come available and indeed the fungi have not yet been made the subject of any investigation in biometrical and population genetics on a major scale. An approach of this kind could hardly fail to be rewarding, and indeed some preliminary observations by my colleague Mr. SIMCHEN already suggest that differences in gene action and interaction may exist between dikaryons from different regions. The fungi, and perhaps especially Basydiomycetes like *Schizophyllum*, offer excellent material for studies of the constitution and inter-relations of populations.

References

GARNJOBST, L.: Genetic control of heterokaryosis in *Neurospora crassa*. Am. J. Botany **40**, 607—614 (1953).

GRINDLE, M.: Heterokaryon compatibility of closely related wild isolates of *Aspergillus nidulans*. Heredity **18**, 397—405 (1963).

JINKS, J.L.: Naturally occuring cytoplasmic changes in fungi. Compt. Rend. Trav. Lab. Carlsberg, Serie Physiologique **26**, 183—203 (1956).

LEWIS, D.: Comparative incompatability in Angiosperms and Fungi. Advan. Genet. **6**, 235—285 (1954).

MATHER, K., and J.L. JINKS: Cytoplasm in sexual reproduction. Nature **182**, 1188 (1958).

RAPER, J.R., G.S. KRONGELB, and M.G. BAXTER: The number and distribution of incompatibility factors in *Schizophyllum*. Am. Naturalist **92**, 221—232 (1958).

RIZET, G.: Les modifications qui conduisent á la sénescence chez *Podospora*, sont-elles de nature cytoplasmique? Compt. Rend. Acad. Sci. (Paris) **244**, 663—665 (1957).

Author Index

Numbers in italics indicate the pages on which the complete references are given.

General Index